# McGraw-Hill
# My Math

CCSS

# Interactive Guide
## Teacher Edition

Grade 4

Mc
Graw
Hill
Education

ConnectED.mcgraw-hill.com

**STEM** McGraw-Hill is committed to providing
instructional materials in Science, Technology,
Engineering, and Mathematics (STEM) that give all
students a solid foundation, one that prepares them for
college and careers in the 21st century.

Send all inquiries to:
McGraw-Hill Education
8787 Orion Place
Columbus, OH 43240

Selections from:
ISBN: 978-0-02-130755-5 *(Grade 4 Student Edition)*
MHID: 0-02-130755-5 *(Grade 4 Student Edition)*
ISBN: 978-0-02-131406-5 *(Grade 4 Teacher Edition)*
MHID: 0-02-131406-3 *(Grade 4 Teacher Edition)*

Printed in the United States of America.

Visual Kinesthetic Vocabulary® is a registered
trademark of Dinah-Might Adventures, LP.

4 5 6 7 8 9 LOV 22 21 20 19 18

# Contents

## Chapter 5 Multiply with Two-Digit Numbers

## Chapter 6 Divide by a One-Digit Number

## Chapter 7 Patterns and Sequences

## Chapter 8 Fractions

# Chapter 13 Perimeter and Area

# Chapter 14 Geometry

# Proficiency Level Descriptors

| | Interpretive (Input) | | Productive (Output) | |
|---|---|---|---|---|
| | **Listening** | **Reading** | **Writing** | **Speaking** |
| **An Emerging Level EL**<br><br>• New to this country; may have memorized some everyday phrases like, "Where is the bathroom", "My name is....", may also be in the "silent stage" where they listen to the language but are not comfortable speaking aloud<br>• Struggles to understand simple conversations<br>• Can follow simple classroom directions when overtly demonstrated by the instructor | • Listens actively yet struggles to understand simple conversations<br>• Possibly understands "chunks" of language; may not be able to produce language verbally | • Reads familiar patterned text<br>• Can transfer Spanish decoding somewhat easily to make basic reading in English seem somewhat fluent; comprehension is weak | • Writes labels and word lists, copies patterned sentences or sentence frames, one- or two-word responses | • Responds non-verbally by pointing, nodding, gesturing, drawing<br>• May respond with yes/no, short phrases, or simple memorized sentences<br>• Struggles with non-transferable pronunciations. |
| **An Expanding Level EL**<br><br>• Is dependent on prior knowledge, visual cues, topic familiarity, and pretaught math-related vocabulary<br>• Solves word problems with significant support<br>• May procedurally solve problems with a limited understanding of the math concept. | • Has ability to understand and distinguish simple details and concepts of familiar/ previous learned topics | • Recognizes obvious cognates<br>• Pronounces most English words correctly, reading slowly and in short phrases<br>• Still relies on visual cues and peer or teacher assistance | • Produces writing that consists of short, simple sentences loosely connected with limited use of cohesive devices<br>• Uses undetailed descriptions with difficulty expressing abstract concepts | • Uses simple sentence structure and simple tenses<br>• Prefers to speak in present tense. |
| **A Bridging Level EL**<br><br>• May struggle with conditional structure of word problems<br>• Participates in social conversations needing very little contextual support<br>• Can mentor other ELs in collaborative activities. | • Usually understands longer, more elaborated directions, conversations, and discussions on familiar and some unfamiliar topics<br>• May struggle with pronoun usage | • Reads with fluency, and is able to apply basic and higher-order comprehension skills when reading grade-appropriate text | • Is able to engage in writing assignments in content area instruction with scaffolded support<br>• Has a grasp of basic verbs, tenses, grammar features, and sentence patterns | • Participates in most academic discussions on familiar topics, with some pauses to restate, repeat, or search for words and phrases to clarify meaning. |

# Strategies for EL Success

Surprisingly, content instruction is one of the most effective methods of acquiring fluency in a second language. When content is the learner's focus, the language used to perform the skill is not consciously considered. The learner is thinking about the situation, or how to solve the problem, not about the grammatical structure of their thoughts. Attempting skills in the target language forces the language into the subconscious mind, where useable language is stored. A dramatic increase in language integration occurs when multiple senses are involved, which causes heightened excitement, and a greater student investment in the situation's outcome. Given this, a few strategies to employ during EL instruction that can make teaching easier and learning more efficient are listed below:

- Activate EL prior knowledge and cultural perspective
- Use manipulatives, realia, and hands-on activities
- Identify cognates
- Build a Word Wall
- Modeled talk
- Choral responses
- Echo reading
- Provide sentence frames for students to use
- Create classroom anchor charts
- Utilize translation tools (i.e. Glossary, eGlossary, online translation tools)
- Anticipate common language problems

## Common Problems for English Learners

| | Cantonese | Haitian Creole | Hmong | Korean | Spanish | Vietnamese |
|---|---|---|---|---|---|---|
| **Phonics Transfers** | | | | | | |
| Pronouncing the /k/as in cake | ● | | ● | ● | | |
| Pronouncing the digraph /sh/ | ● | | ● | | ● | ● |
| Hearing and reproducing the /r/, as in *rope* | ● | | ● | ● | | ● |
| /j/ | | | ● | | ● | |
| Hearing or reproducing the short /u/ | | ● | ● | | | |
| **Grammar Transfers** | | | | | | |
| Adjectives often follow nouns | | ● | ● | | ● | ● |
| Adjectives and adverb forms are interchangeable | | ● | ● | | | |
| Nouns have feminine or masculine gender | | | | | ● | |
| There is no article or there is no difference between articles *the* and *a* | | ● | ● | | | ● |
| Shows comparative and superlative forms with separate words | | | ● | | ● | |
| There are no phrasal verbs | | | | ● | ● | |

# How to Use the Teacher Edition

The Interactive Guide provides scaffolding strategies and tips to strengthen the quality of mathematics instruction. The suggested strategies, activities, and tips provide additional language and concept support to accelerate English learners' acquisition of academic English.

## English Learner Instructional Strategy

Each lesson – including Inquiry/Hands On and Problem-Solving Investigation – references an English Learner Instructional Strategy that can be utilized before or during regular class instruction. These strategies specifically support the Teacher Edition and scaffold the lesson for English learners (ELs).

Categories of the scaffolded support are:
- Vocabulary Support
- Language Structure Support
- Sensory Support
- Graphic Support
- Collaborative Support

The goal of the scaffolding strategies is to make each individual lesson more comprehensible for ELs by providing visual, contextual and linguistic support to foster students' understanding of basic communication in an academic context.

### Lesson 2 Estimate Quotients
*English Learner Instructional Strategy*

**Sensory Support: Illustrations, Diagrams, Drawings**

Before the lesson, review the terms *compatible numbers, fact family, basic fact,* and *place value.* Write the terms on chart paper, and ask students to generate a definition and math example for you to write next to each term. Also invite students to suggest any illustrations, diagrams, or drawings that help them remember the meanings of the terms, and note them beside the terms' definitions. Encourage students to copy this information in their math journals.

To help students discuss the Talk Math problem, provide this communication guide:
**To estimate $\$4,782 \div 6$, first round $\$4,782$ up to ____.**
**Then divide ____ by 6.**
**The basic fact related to the numbers in this equation is**
**____ × ____ = ____. The estimate is ____.**

Since ELs benefit from visual references to new vocabulary, many of the English Learner Instruction Strategies suggest putting vocabulary words as well as Spanish cognates on a Word Wall. Choose a location in your classroom for your Word Wall, and organize the words by chapter, by topic, by Common Core domain, or alphabetically.

# How to Use the Teacher Edition *continued*

## English Language Development Leveled Activities

These activities are tiered for Emerging, Expanding, and Bridging leveled ELs. Activity suggestions are specific to the content of the lesson. Some activities include instruction to support students with lesson-specific vocabulary they will need to understand the math content in English, while other activities teach the concept or skill using scaffolded approaches specific to ELs. The activities are intended for small group instruction, and can be directed by the instructor, an aide, or a peer mentor.

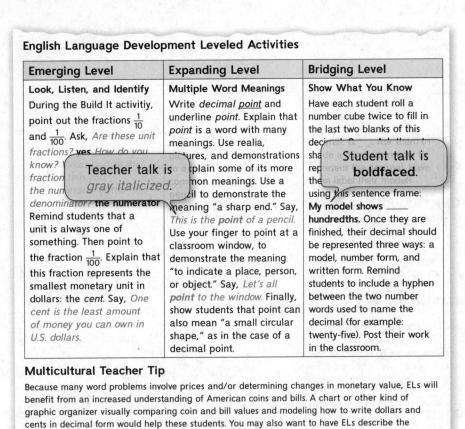

**English Language Development Leveled Activities**

| Emerging Level | Expanding Level | Bridging Level |
|---|---|---|
| **Look, Listen, and Identify**<br>During the Build It activitiy, point out the fractions $\frac{1}{10}$ and $\frac{1}{100}$. Ask, *Are these unit fractions? yes How do you know? ... fraction ... the num... denominator?* **the numerator** Remind students that a unit is always one of something. Then point to the fraction $\frac{1}{100}$. Explain that this fraction represents the smallest monetary unit in dollars: the *cent*. Say, *One cent is the least amount of money you can own in U.S. dollars.* | **Multiple Word Meanings**<br>Write *decimal point* and underline *point*. Explain that *point* is a word with many meanings. Use realia, pictures, and demonstrations to explain some of its more common meanings. Use a pencil to demonstrate the meaning "a sharp end." Say, *This is the point of a pencil.* Use your finger to point at a classroom window, to demonstrate the meaning "to indicate a place, person, or object." Say, *Let's all point to the window.* Finally, show students that point can also mean "a small circular shape," as in the case of a decimal point. | **Show What You Know**<br>Have each student roll a number cube twice to fill in the last two blanks of this decimal 0.__. Ask them to shade ... represent ... have them label their models using this sentence frame: **My model shows ___ hundredths.** Once they are finished, their decimal should be represented three ways: a model, number form, and written form. Remind students to include a hyphen between the two number words used to name the decimal (for example: twenty-five). Post their work in the classroom. |

*Teacher talk is gray italicized.*

**Student talk is boldfaced.**

**Multicultural Teacher Tip**

Because many word problems involve prices and/or determining changes in monetary value, ELs will benefit from an increased understanding of American coins and bills. A chart or other kind of graphic organizer visually comparing coin and bill values and modeling how to write dollars and cents in decimal form would help these students. You may also want to have ELs describe the monetary systems of their native countries. Identifying similarities or differences with the American system can help familiarize students with dollars and cents.

## Multicultural Teacher Tip

These tips provide insight on academic and cultural differences you may encounter in your classroom. While math is the universal language, some ELs may have been shown different methods to find the answer based on their native country, while cultural customs may influence learning styles and behavior in the classroom.

# What's the Math in this Chapter?

## Mathematical Practice

The goals of the Mathematical Practice activity are to help students clarify the specific language of the Mathematical Practice and rewrite the Mathematical Practice in simplified language that students can **relate** to. Examples are discussed to make connections, showing students how they use this Mathematical Practice to solve math problems.

### Chapter 2 Add and Subtract Whole Numbers

*What's the Math in This Chapter?*

**Mathematical Practice 8: Look for and express regularity in repeated reasoning.**

Prepare a demonstration place-value chart with the following numbers written: 536; 546; 556. Leave a black space below 556. Display the chart and say, *Look at the chart carefully. Do you see a pattern?*

Allow students time to think on their own. Then encourage them to turn and talk with a peer. Discuss as a group. The discussion goal is for students to identify the "add ten" pattern. Ask, *Now that we know the pattern is to add ten, what is the next number?* **566** Write 566 on the place-value chart.

Discuss with students the pattern of adding 10 is a repeated calculation. Say, *When you figure*

*repeated reasoning.*

> Mathematical Practice is rewritten as an "I can" statement.

Display a chart with Mathematical Practice 8. Relate Mathematical Practice 8 and have students assist in rewriting it as an "I can" statement, for example: **I can solve** problems by using repeated calculations. Have students draw or write examples of using repeated reasoning/repeated calculations. Post examples and the new "I can" statement.

*Inquiry of the Essential Question:*

**What strategies can I use to add or subtract?**

Inquiry Activity Target: **Students come to a conclusion that they can use repeated calculations to solve problems.**

> Inquiry Activity Target connects Mathematical Practice to Essential Question.

As an introduction to the chapter, present the Essential Question to graphic organizer will offer opportunities for students inferences, and apply prior knowledge of repeated questions representing the Essential Question. As they investigate, students to draw, write, and collaborate with peers to demonstrate their observations and thinking. Then have students present additional questions they may have to a peer to extend discussions.

## Inquiry of the Essential Question

As an introduction to the Chapter, the Inquiry of the Essential Question graphic organizer activity is designed to introduce the Essential Question. The activity offers opportunities for students to observe, make inferences, and apply prior knowledge of samples/models representing the Essential Question. Collaborative conversations drive students toward the Inquiry Activity Target which is to make a connection between the "Mathematical Practice of the chapter" and the "Essential Question of the chapter."

# How to Use the Student Edition

Each student page provides EL support for vocabulary, note taking, and speaking/writing skills. These pages can be used before, during, or after each classroom lesson. A corresponding page with answers is found in the Teacher Edition.

## Inquiry of the Essential Question

Students observe, make inferences, and apply prior knowledge of chapter specific samples/models representing the Essential Question of the chapter. Encourage students to have collaborative conversations as they share their ideas and questions with peers. As students inquire the math models, present specific questions that will drive students toward the Inquiry Activity Target which is stated on the Teacher Edition page.

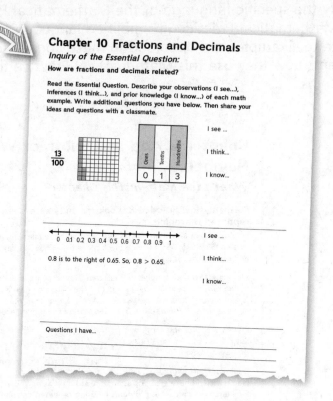

### Chapter 10 Fractions and Decimals

*Inquiry of the Essential Question:*

**How are fractions and decimals related?**

Read the Essential Question. Describe your observations (I see...), inferences (I think...), and prior knowledge (I know...) of each math example. Write additional questions you have below. Then share your ideas and questions with a classmate.

$\frac{13}{100}$

| Ones | Tenths | Hundredths |
|------|--------|------------|
| 0 | 1 | 3 |

I see ...

I think...

I know...

0.8 is to the right of 0.65. So, 0.8 > 0.65.

I see ...

I think...

I know...

Questions I have...

## Cornell Notes/Note Taking

Cornell notes offer students a method to use to take notes, thereby helping them with language structure. Scaffolded sentence frames are provided for students to fill in important math vocabulary by identifying the correct word or phrase according to context. Encourage students to refer to their books to locate the words needed to complete the sentences. Each note taking graphic organizer will support students in answering the Building on the Essential Question.

### Lesson 5 Note Taking

*Relate Area and Perimeter*

Read the question. Write words you need help with and research each word. Use your lesson to write your Cornell notes. Write or draw math examples to explain your thinking. Share your examples with a classmate.

| Building on the Essential Question | Notes: |
|---|---|
| How are area and perimeter related? | Area is the number of _____ _____ needed to cover the _____ of a region or plane figure without any _____.  Perimeter is the _____ a shape or region.  Two figures _____ have the same _____ and different areas. |
| Words I need help with: | Perimeter: ___ units    Perimeter: ___ units  Area: ___ square units    Area: ___ square units  Two figures _____ have the same _____ and different perimeters.  Perimeter: ___ units    Perimeter: ___ units  Area: ___ square units    Area: ___ square units |

## Vocabulary Cognates

Students define each vocabulary word or phrase and write a sentence using the term in context. Space is provided for Spanish speakers to write the definition in Spanish. A blank Vocabulary Word Boxes template is provided on page xx in this book for use with non-Spanish speaking ELs.

### Lesson 2 Vocabulary Cognates
*Read and Write Multi-Digit Numbers*

Use the Glossary to define the math word in English and in Spanish in the word boxes. Write a sentence using your math word.

| expanded form | forma desarrollada |
|---|---|
| Definition | Definición |
| My math word sentence: | |

| word form | forma verbal |
|---|---|
| Definition | Definición |
| My math word sentence: | |

| standard form | forma estándar |
|---|---|
| Definition | Definición |
| My math word sentence: | |

### Lesson 3 Guided Writing
*Inquiry/Hands On: Use Place Value to Multiply*

**How do you use place value to multiply?**

Use the exercises below to help you build on answering the Essential Question. Write the correct word or phrase on the lines provided.

1. Rewrite the question in your own words.

2. What key words do you see in the question?

3. Identify the number that is modeled below.

4. To model the multiplication equation 2 × 41, you would model the number _____ using base ten blocks _____ times.

5. What multiplication **equation** is modeled below?

6. Use the modeled multiplication equation above to identify the total number of tens and ones. Record in the place-value chart.

| tens | ones |
|---|---|

7. How do you use place value to multiply?

## Guided Writing

Guided writing provides support to help ELs meet the stated Lesson Objective. Content specific questions are scaffolded to build language knowledge in order to answer the question. Give bridging level students the opportunity to mentor and assist emerging and expanding level students when answering the questions, if needed.

# How to Use the Student Edition *continued*

## Vocabulary Chart

Three-column charts concentrate on English/ Spanish cognates. Students are given the word in English. Encourage students to use the Glossary to find the word in Spanish and the definition in English. As an extension, have students identify and highlight other cognates which may be in the definitions. A blank Vocabulary Chart template is provided on page xix in this book for use with non-Spanish speaking ELs.

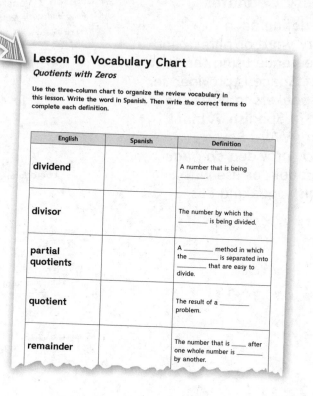

**Lesson 10 Vocabulary Chart**
*Quotients with Zeros*

Use the three-column chart to organize the review vocabulary in this lesson. Write the word in Spanish. Then write the correct terms to complete each definition.

| English | Spanish | Definition |
|---------|---------|------------|
| dividend | | A number that is being _____ . |
| divisor | | The number by which the _____ is being divided. |
| partial quotients | | A _____ method in which the _____ is separated into _____ that are easy to divide. |
| quotient | | The result of a _____ problem. |
| remainder | | The number that is _____ after one whole number is _____ by another. |

**Lesson 2 Concept Web**
*Numeric Patterns*

Use the given rule to write the next number in each numeric pattern shown in the concept web.

2, 8, 14, _____

1, 7, 13, _____

3, 9, 15, _____

The rule is +6.

## Concept Web

Concept webs are designed to show relationships between concepts and to make connections. As each concept web is unique in design, please read and clarify directions for students. Encourage students to look through the lesson pages to find examples or words they can use to complete the web.

## Definition Map

The definition maps are designed to address a single vocabulary word, phrase, or concept. Students should use the Glossary to help define the word in the description box. Sentence frames are provided to scaffold characteristics from the lesson. Students can refer to the lesson examples and the Glossary to assist them in completing the sentence frames as well as creating their own math examples. Make sure you review with students the tasks required.

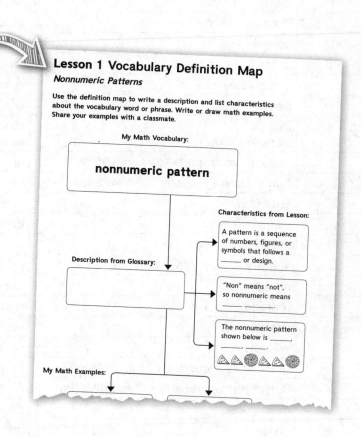

**Lesson 1 Vocabulary Definition Map**
*Nonnumeric Patterns*

Use the definition map to write a description and list characteristics about the vocabulary word or phrase. Write or draw math examples. Share your examples with a classmate.

My Math Vocabulary:

**nonnumeric pattern**

Characteristics from Lesson:

A pattern is a sequence of numbers, figures, or symbols that follows a _____ or design.

"Non" means "not", so nonnumeric means _____.

The nonnumeric pattern shown below is _____.

Description from Glossary:

My Math Examples:

---

**Lesson 6 Problem-Solving Investigation**
*STRATEGY: Make a Table*

Make a table to solve each problem.

1. A page from Dana's album is shown. Dana puts the <u>same</u> number of stickers on <u>each</u> page. <u>She</u> (Dana) has **30 pages** of stickers. **How many** stickers does she have <u>in all</u>?

| Understand | Solve |
|---|---|
| I know: | Pages: 1, 10, 20, 30 / Stickers: 12 |
| I need to find: | Dana has _____ stickers. |
| Plan | Check |
| I will make a _____ to solve. | |

2. West Glenn School has **23 students** in <u>each</u> **class**. There are **6** fourth grade **classes**. **About** how many fourth grade **students** are there <u>in all</u>?

| Understand | Solve |
|---|---|
| I know: | Classes: 1, 2, 3, 4, 5, 6 / Students: 20 |
| I need to find: | There are **about** _____ students. |
| Plan | Check |
| I will make a _____ to solve. | |

## Problem-Solving Investigation

Each Problem-Solving Investigation page focuses on scaffolding Exercises 1 and 2 from the Apply the Strategy portion of the lesson in the book. The text for each exercise highlights signal words and phrases to help students decipher the key information in the problem. Visual images as well as tables and sentence frames are included to assist students with the problem-solving process. A blank Problem-Solving Investigation template is provided on page xxii in this book if students need additional assistance with other exercises.

# English/Spanish Cognates Used in Grade 4

| Chapter | English | Spanish |
|---------|---------|---------|
| 1 | digit | dígito |
| | is equal to (=) | es igual a (=) |
| | number line | recta numérica |
| | order | ordenar (verb) orden (noun) |
| | place value | valor posicional |
| | period | período |
| | round<br>standard form | redondear forma<br>estándar |
| 2 | Associative Property (of Addition) | propiedad asociativa (de la suma) |
| | Commutative Property (of Addition) | propiedad conmutativa (de la suma) |
| | equation | ecuación |
| | estimate | estimación (noun) estimar (verb) |
| | difference | diferencia |
| | Identity Property (of Addition) | propiedad identidad (de la suma) |
| | minuend | minuendo |
| | regroup | reagrupar |
| | subtrahend | sustraendo |
| | variable | variable |
| 3 | Associative Property of Multiplication | propiedad asociativa de la multiplicación |
| | Commutative Property of Multiplication | propiedad conmutativa de la multiplicación |
| | decompose | descomponer |
| | divide | dividir |
| | dividend | dividendo |
| | division | división |
| | divisor | divisor |
| | factors | factores |
| | Identity Property of Multiplication | propiedad de identidad de la multiplicación |
| | multiple<br>multiply | múltiplo<br>multiplicar |
| | multiplication | multiplicación |
| | quotient | cociente |
| | product | producto |
| | Zero Property of Multiplication | propiedad del cero de la multiplicación |
| 4 | Distributive Property | propiedad distributiva |
| | partial products | productos parciales |
| 5 | operation | operación |
| 6 | compatible numbers | números compatibles |
| | parentheses | paréntesis |
| | partial quotients | cocientes parciales |
| 7 | nonnumeric pattern | patrón no numerico |

| Chapter | English | Spanish |
|---|---|---|
| | numeric pattern | patrón numeric |
| | order of operations | orden de las operaciones |
| | sequence | secuencia |
| | term | término |
| 8 | composite number | número compuesto |
| | denominator | denominador |
| | equivalent fractions | fracciones equivalentes |
| | factor pairs | pares de factores |
| | fraction | fracción |
| | (greatest) common factor | (máximo) factor común |
| | improper fraction | fracción impropia |
| | (least) common multiple | (mínimo) común múltiplo |
| | mixed number | número mixto |
| | numerator | numerador |
| | prime number | número primo |
| 9 | improper fraction | fracción impropia |
| | simplify | simplificar |
| | unit fraction | fracción unitaria |
| 10 | decimal | decimal |
| 11 | capacity | capacidad |
| | convert | convertir |
| | gallon (gal) | galón (gal) |
| | mile (mi) | milla (mi) |
| | ounce (oz) | onza (oz) |
| | pint (pt) | pinta (pt) |
| | quart (qt) | cuarto (ct) |
| | seconds | segundos |
| | ton (T) | tonelada (T) |
| | yard | yarda |
| 12 | centimeter (cm) | centímetro (cm) |
| | kilometer (km) | kilómetro (km) |
| | meter (m) | metro (m) |
| | metric system | sistema metric (SI) |
| | millimeter (mm) | milímetro (mm) |
| | liter (L) | litro (L) |
| | milliliter(mL) | mililitro (mL) |
| | gram (g) | gramo (g) |
| | kilogram (kg) | kilogramo (kg) |
| | mass | masa |

# English/Spanish Cognates Used in Grade 4 *continued*

| Chapter | English | Spanish |
|---|---|---|
| 13 | perimeter | perímetro |
| | area | área |
| 14 | acute angle | ángulo agudo |
| | acute triangle | triágulo agudo |
| | angle | ángulo |
| | degree (°) | grado (°) |
| | line | recta |
| | line symmetry | simetria axial |
| | line segment | segemento de recta |
| | obtuse angle | ángulo obtuso |
| | obtuse triangle | triágulo obtuso |
| | parallel (ll) | paralela |
| | parallelogram | paralelogramo |
| | perpendicular | perpendicular |
| | rectangle | rectángulo |
| | rhombus | rombo |
| | right angle | ángulo recto |
| | right triangle | triágulo recto |
| | trapezoid | trapecio |

# Vocabulary Chart
## Chapter _____, Lesson _____

Use the three-column chart to organize the vocabulary in this lesson.
Write the word in your own language. Then write each definition.

| English | Native Language | Definition |
|---|---|---|
|  |  |  |
|  |  |  |
|  |  |  |
|  |  |  |
|  |  |  |
|  |  |  |

# Vocabulary Word Boxes

## Chapter _____, Lesson _____

Use the word boxes to define the math word in English and in your
native language. Write a sentence using your math word.

| | |
|---|---|
| | |
| **Definition** | |

**My math word sentence:**

| | |
|---|---|
| | |
| **Definition** | |

**My math word sentence:**

# Vocabulary Definition Map
## Chapter _____, Lesson _____

Use the definition map to write a description and list characteristics about the vocabulary word or phrase. Write or draw math examples. Share your examples with a classmate.

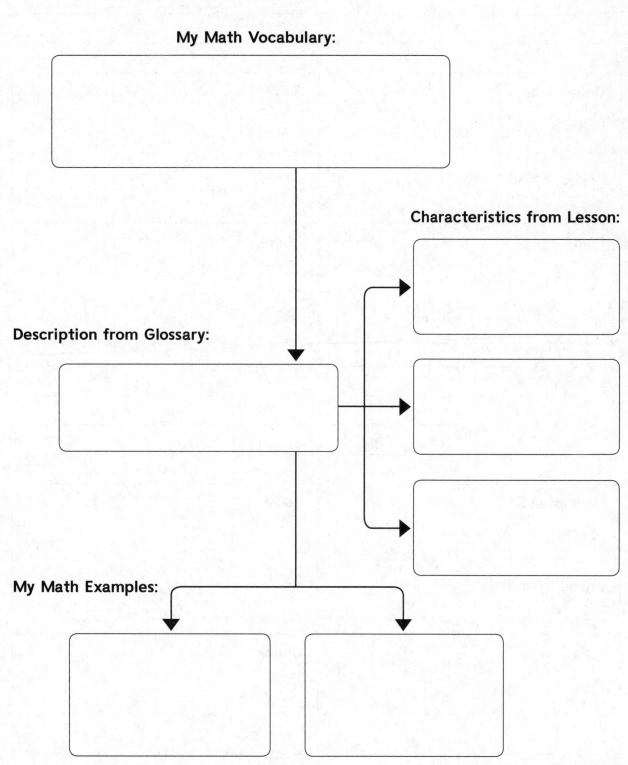

My Math Vocabulary:

Characteristics from Lesson:

Description from Glossary:

My Math Examples:

# Problem-Solving Investigation
## Chapter _____, Lesson _____

1. _____

_____

_____

_____

| Understand | Solve |
|---|---|
| I know:<br><br>I need to find: | |
| **Plan** | **Check** |

2. _____

_____

_____

_____

| Understand | Solve |
|---|---|
| I know:<br><br>I need to find: | |
| **Plan** | **Check** |

# Chapter 1 Place Value

## What's the Math in This Chapter?

### Mathematical Practice 1: Make sense of problems and persevere in solving them.

Write and read the following word problem. *Oliver has 241 shells he has collected. His brother, Max has collected 126 shells. Oliver and Max have decided to combine their collections. How many shells do they have in all?*

Discuss with students what they know and what they need to find in this problem. Ask, *How should we solve this problem?* **add the numbers together** On the board write 241 + 126 vertically but do not line the numbers up correctly. Put the 1 in 126 under the 4 in 241 instead of the 2. Keep solving the problem, even if students are protesting.

Say, *Okay, I solved the problem. Oliver and Max have 2,536 shells. Wait! Does 2,536 shells* **make sense***?* Allow students time to think about the way you solved the problem. When a student identifies that you didn't add correctly say, *Oh you are right. I* **understand** *what I did wrong. I didn't line up the numbers correctly.*

Draw a place-value chart and write the numbers correctly aligned in the chart. Model proper addition showing the sum to be 367. Discuss with students how using place value helps us to **make sense of problems** and add correctly.

Display a chart with Mathematical Practice 1. Restate Mathematical Practice 1 and have students assist in rewriting it as an "I can" statement, for example: **I can better understand problems by using place value.** Post the new "I can" statement.

## Inquiry of the Essential Question:

### How does place value help represent the value of numbers?

Inquiry Activity Target: **Students come to a conclusion that understanding place value is important in solving problems.**

As an introduction to the chapter, present the Essential Question to students. The inquiry graphic organizer will offer opportunities for students to observe, make inferences, and apply prior knowledge of problem solving representing the Essential Question. As they investigate, encourage students to draw, write, and collaborate with peers to demonstrate their observations and thinking. Then have students present additional questions they may have to a peer to extend discussions.

Regroup students and restate Mathematical Practice 1 and the Essential Question. Pose questions to reflect on what has been learned to guide students in making connections between the Mathematical Practice and the Essential Question.

NAME _____ DATE _____

# Chapter 1 Place Value

## *Inquiry of the Essential Question:*

**How does place value help represent the value of numbers?**

Read the Essential Question. Describe your observations (I see...),
inferences (I think...), and prior knowledge (I know...) of each math
example. Write additional questions you have below. Then share your
ideas and questions with a classmate.

| Thousands | | | Ones | | |
|---|---|---|---|---|---|
| hundreds | tens | ones | hundreds | tens | ones |
| 6 | 2 | 8 | 1 | 5 | 0 |

The value of 6 is 6 × 100,000 or 600,000.

I see ...

I think...

I know...

---

**Standard Form:** 931,057

**Expanded Form:** 900,000 + 30,000 + 1,000 +
50 + 7

**Word Form:** *nine hundred thirty-one thousand,*
*fifty-seven*

I see ...

I think...

I know...

---

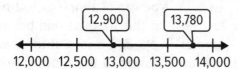

12,000   12,500   13,000   13,500   14,000

I see ...

I think...

I know...

---

Questions I have...

_____

_____

_____

# Lesson 1 Place Value

## English Learner Instructional Strategy

### Vocabulary Support: Utilize Resources

Before the lesson, write *digit* and its Spanish cognate, *dígito*, on the board. Introduce the words, and provide concrete examples (realia) to support understanding. Utilize other appropriate translation tools for non-Spanish speaking ELs.

Have a bilingual aid or a peer review the components of a place-value chart. During the lesson, point to the corresponding place-value positions on a demonstration chart as you discuss the lesson examples.

For Talk Math, pair emerging students with expanding/bridging level bilingual peers. Provide this sentence frame for students to report back: **The value is ____ times greater in the ____ place.**

### English Language Development Leveled Activities

| Emerging Level | Expanding Level | Bridging Level |
|---|---|---|
| **Developing Oral Language** | **Act It Out** | **Number Game** |
| Display a three-column chart with these headings: **Column 1**: *hundreds* **Column 2**: *tens* **Column 3**: *ones* Then write 563 beside it. Say, *We read numbers from left to right, but we look at place value from right to left.* Rewrite 5 in Column 1, 6 in Column 2, and 3 in Column 3. Then write this sentence frame on the board: **The digit ____ is in the ____ place.** Point to each digit and tell its value. For example, point to 3 and say, *The digit 3 is in the ones place.* Then have students chorally repeat. | Divide students into groups of three, and distribute one number cube to each group. Assign one student each to represent the ones, tens, and hundreds place values. Have each student in the group roll the number cube. Then have students put their digits together to create a three-digit number based on each student's assigned place value. Have each group share its number with the class using the following sentence frame: **The digit ____ is in the ____ place.** | Write a six-digit number on the board. Then have two students come to the board. Assign each student one digit from the number. Then have students race to write the value of the digit as a number multiplied by a power of ten. (For example, if the six-digit number is *413,268* and a student is assigned the digit 3, the student would write $3 \times 1,000 = 3,000$.) Have each student say aloud his or her answer. |

**Teacher Notes:**

NAME _____ DATE _____

# Lesson 1 Four-Square Vocabulary
## *Place Value*

Write the definition for each math word. Write what each word means in your own words. Draw or write examples that show each math word meaning. Then write your own sentences using the words.

| Definition | My Own Words |
|---|---|
| A symbol used to write numbers. | See students' examples. |

**digit**

| My Examples | My Sentence |
|---|---|
| Sample answer: The ten digits are 0, 1, 2, 3, 4, 5, 6, 7, 8, and 9. | Sample sentence: The number 459 contains three digits: 4, 5, and 9. |

| Definition | My Own Words |
|---|---|
| The value given to a *digit* by its position in a number. | See students' examples. |

**place value**

| Thousands Period | | | Ones Period | | |
|---|---|---|---|---|---|
| hundreds | tens | ones | hundreds | tens | ones |
| 2 | 8 | 4 | 7 | 5 | 1 |

2,

| My Examples | My Sentence |
|---|---|
| Sample answer: The 5 is in the tens place. | Sample sentence: The digit in the ten thousands place has the value 80,000. |

# Lesson 2 Read and Write Multi-Digit Numbers

## English Learner Instructional Strategy

### Cooperative Learning: Partners Work/Pairs Check

Before the lesson, write *standard form* and its Spanish cognate, *forma estándar*, on a cognate chart. Introduce the words and write a math example to support understanding.

Assign Problem Solving Exercises 14–18. Have students work in pairs. One student completes the first problem while the second acts as a coach. Then, students switch roles for the second problem. When they finish the second problem, they get together with another pair and check answers. Provide the following sentence frames: **What answer did you get for Exercise _____? Our answer is _____.** When both pairs have agreed on the answers, ask them to shake hands and continue working in original pairs on the next two problems. Partners work exchanging roles as coach, and then new pairs check answers.

## English Language Development Leveled Activities

| Emerging Level | Expanding Level | Bridging Level |
|---|---|---|
| **Listen and Identify** Display these items: 1. *8,657* 2. *eight thousand, six hundred fifty-seven* 3. *8,000 + 600 + 50 + 7*. Point to Item 1 and say, *standard form*. Have students repeat chorally. Explain that standard form uses numbers only. Point to Item 2 and say, *word form*. Have students repeat chorally. Explain that word form uses words only. Point to Item 3 and say, *expanded form*. Have students repeat chorally. Explain that expanded form shows the value of each digit. Then randomly point to each of the three items and have students verbally name its form. | **Building Oral Language** Prior to class, prepare 8–10 sheets of paper on which you have written numbers in either standard, expanded, or word form. Then display each paper and have students identify the number's form using the following sentence frame: **The number is in _____ form.** To extend the activity, have students write each number you display in at least one other form. | **Number Recognition** On slips of paper, write five-digit or six-digit numbers in word form. Put the slips of paper in a basket or box. Then have each student choose a slip of paper and read the number. Have each student write his or her number in both standard form and expanded form. |

**Teacher Notes:**

NAME _____ DATE _____

# Lesson 2 Vocabulary Cognates
## *Read and Write Multi-Digit Numbers*

Use the Glossary to define the math word in English and in Spanish
in the word boxes. Write a sentence using your math word.

| expanded form | forma desarrollada |
|---|---|
| **Definition** <br> The representation of a number as a *sum* that shows the value of each digit. | **Definición** <br> Representación de un número como la *suma* del valor de cada dígito. |

**My math word sentence:**
Sample answer: In expanded form, 423 is 400 + 20 + 3.

| word form | forma verbal |
|---|---|
| **Definition** <br> The form of a number that uses written words. | **Definición** <br> Forma de expresar un número usando palabras escritas. |

**My math word sentence:**
Sample answer: In word form, 168 is one hundred sixty-eight.

| standard form | forma estándar |
|---|---|
| **Definition** <br> The usual way of writing a number that shows only its *digits*, no words. | **Definición** <br> Manera habitual de escribir un número usando solo sus *dígitos*, sin usar palabras. |

**My math word sentence:**
Sample answer: In standard form, three hundred seventy-five is 375.

Grade 4 • **Chapter 1** *Place Value* **3**

# Lesson 3 Compare Numbers

## English Learner Instructional Strategy

### Vocabulary Support: Signal Words/Phrases

Before the lesson, write *number line* and *equal* and their Spanish cognates, *línea numérica* and *igual*, respectively, on a cognate chart. Introduce the words and provide a concrete example. Utilize other appropriate translation tools for non-Spanish speaking ELs.

Introduce the concepts *less than*, *equal to*, and *greater than*. Write the following synonym pairs on the board: *less/fewer, equal/same, greater/more*. Discuss their meanings. Then divide students into two groups: Red and Blue. Assign more students to Red than Blue, and ask group members to stand together. Say, *Red Group has more people than Blue Group. Is the number of people in Red Group greater than or less than the number of people in Blue group?* **Greater than** Then say, *Blue Group has fewer members than Red Group. Is the number of people in Blue Group greater than or less than Red Group?* **Less than** Discuss how to make the two groups equal, or the same.

### English Language Development Leveled Activities

| Emerging Level | Expanding Level | Bridging Level |
|---|---|---|
| **Listen and Identify** | **Number Sense** | **Number Game** |
| Display these symbols: <, >, =. Point to < and say, *This symbol means "is less than."* Point to > and say, *This symbol means "is greater than."* Point to = and say, *This symbol means "is equal to."* Then randomly point to each symbol, and have students chorally say the relationship it represents. Provide the following sentence frame: **This symbol means _____.** | Display a number line from 100 to 900 with increments of 100. Point to it and say, *This is a number line.* Have students repeat chorally. Explain that on the number line, numbers to the right are greater than numbers on the left, and numbers to the left are less than numbers on the right. Then write *450* and *230* on the board. Use the number line to determine their relationship. Then write > between the numbers and say, *450 is greater than 230.* Repeat this activity with students, using other pairs of three-digit numbers. | Divide students into pairs. Distribute two number cubes to each pair, and have each student roll a number cube 4 to 6 times to create a multi-digit number. Then have pairs look at the two multi-digit numbers they created and write two true sentences about how the numbers relate to each other. Tell students to use one of the following symbols in each sentence: <, >, =. (For example: **478,920 > 2,376.**) Then have partners say aloud their number sentences. (For example: **478,920 is greater than 2,376.**) |

### Multicultural Teacher Tip

ELs who are familiar with the US standard for placing the angle symbol at the front of a number or angle designation may be confused by the use of inequality symbols. They may have trouble distinguishing between the two signs, so it is important to emphasize the difference prior to beginning the lesson.

NAME _____ DATE _____

# Lesson 3 Vocabulary Chart
## *Compare Numbers*

Use the three-column chart to organize the vocabulary in this lesson. Write the word in Spanish. Then write the correct terms to complete each definition.

| English | Spanish | Definition |
|---|---|---|
| **number line** | línea numérica | A line with numbers on it in <u>order</u> at regular intervals. |
| **is greater than (>)** | es mayor que (>) | An inequality relationship showing that the number on the <u>left</u> of the symbol is <u>greater</u> than the number on the right. |
| **is less than (<)** | es menor que (<) | An inequality relationship showing that the number on the <u>left</u> side of the symbol is <u>smaller</u> than the number on the right side. |
| **is equal to (=)** | es igual a (=) | Having the <u>same</u> value. The (=) sign is used to show two numbers or expressions are <u>equal</u>. |

# Lesson 4 Order Numbers
## English Learner Instructional Strategy

### Vocabulary Support: Communication Guide

Write *order* and its Spanish cognates, *ordenar* and *orden*, on a cognate chart. Explain the words' meanings. Point out that in English *order* can mean "to command," "to place an order," or "to organize." It can also mean "the way things are arranged." Provide pictures to support each meaning. *In a math context it means "to organize" and "the way things are arranged."*

Create a communication guide with these sets of sentence frames:
1. is the *greater* number. It is more than _____.
2. is the *greatest* number. It is more than _____ and _____.
3. is the *least* number. It is less than _____ and _____.

Explain the chart. Say, *The word **greater** compares two numbers. The word **greatest** compares three or more numbers. The word **least** also compares three or more numbers.* Have students suggest numbers to fill the blanks. Encourage students to write the sentence frames in their math journals and utilize them during lesson discussions.

### English Language Development Leveled Activities

| Emerging Level | Expanding Level | Bridging Level |
|---|---|---|
| **Word Knowledge** Display this number series: *13,440; 13,988; 14,050.* Identify its least and greatest numbers. Then move your hand from left to right along the row of numbers and say, *The numbers are ordered from least to greatest.* Have students repeat, **Least to greatest.** Then display this series: *22,300; 22,066; 21,550.* Identify its greatest and least numbers. Then move your hand from left to right along the row of numbers and say, *The numbers are ordered from greatest to least.* Have students repeat, **Greatest to least.** | **Act It Out** Prior to class prepare a set of index cards in which a different five-digit or six-digit number is written on each card. Then divide students into groups of three or four, and have each student choose a number card. Have the students within each group compare their cards and then stand in order from *least* to *greatest*. Then have students choose different cards to compare. This time have students stand in order from *greatest* to *least*. Repeat the exercise multiple times, using different numbers. | **Public Speaking Norms** Divide students into two multilingual groups. Assign each group a set of five 6-digit numbers. Have one group order their numbers from *least* to *greatest*, and have the other group order their numbers from *greatest* to *least*. Then have each group prepare and present an explanation of the process they used for ordering their numbers. |

### Teacher Notes:

NAME _____ DATE _____

# Lesson 4 Note Taking
## *Order Numbers*

Read the question. Write words you need help with and research each word. Use your lesson to write your Cornell notes. Write or draw math examples to explain your thinking. Share your examples with a classmate.

| Building on the Essential Question | Notes: |
|---|---|
| How does a place value chart help order numbers? | Write the numbers being ordered with each _____digit_____ on a place-value chart. |
| | Start with the greatest _place value_ position. |
| | Compare using the _____greater_____ than symbol (>). |
| | Compare the _____digits_____ in the next _place value_ position. |
| | Continue to compare until the _____digits_____ are different. |

**Words I need help with:**

See students' words

**My Math Examples:**

See students' examples

# Lesson 5 Use Place Value to Round
## *English Learner Instructional Strategy*

### Cooperative Learning: Think-Pair-Share

Explain that the verb phrase to *round* can have multiple meanings in English. For example, it can mean "to gather, to make something round, or to move along a curved path." Provide pictures to support meaning. Then explain that in a math context, to *round* means "to make a number easier to work with." For example, model that 12 rounds to 10, and show that 10 is an easy number to add, subtract, multiply, or divide.

Have partners work together to discuss the Talk Math question. If needed, ask, *Is the "least number" the highest number or lowest number?* **The lowest number** Have students think about what the lowest number is that rounds up to 8,000 and discuss their reasoning with a partner. Then invite volunteers to share and explain their solutions to the problem. Have the whole group vote to show whether they agree or disagree with a solution.

### English Language Development Leveled Activities

| Emerging Level | Expanding Level | Bridging Level |
|---|---|---|
| **Word Knowledge** Say, *We round up to make a number's value more.* Have students repeat chorally, **Round up.** Say, *We round down to make a number's value less.* Have students repeat chorally, **Round down.** Then display a number line numbered 20–30. Say, *The number 27 is closer to 30 than 20. Therefore, 27 will round up to 30. The number 22 is closer to 20 than 30. Therefore, 22 will round down to 20.* Say other numbers between 20 and 30. For each, have students point up if the number rounds up or point down if the number rounds down. | **Internalize Language** Write 341,852 on the board. Say, *Let's round this number to the nearest thousand.* Circle 1 and say, *1 is in the thousands place.* Then underline 1,852 and say, *1,852 is closer to 2,000 than 1,000. Therefore, we will round up 341,852 to 342,000.* Now demonstrate how to round to the nearest ten thousand. Circle 4 and say, *The 4 is in the ten thousands place.* Underline 41,852 and say, *41,852 is closer to 40,000 than 50,000. Therefore, we round down 341,852 to 340,000.* Repeat the exercise with other multi-digit numbers. Encourage student input. | **Show What You Know** Divide students into multilingual groups, and distribute a number cube to each group. Have each group roll their number cube six times and use each number rolled to create a six-digit number. Then have each group round their six-digit number to the nearest ten, hundred, thousand, ten thousand, and hundred thousand. Direct students to present their original number and rounded numbers to the class, explaining their reasoning for each rounded number. |

### Teacher Notes:

NAME _____ DATE _____

# Lesson 5 Concept Web

## *Use Place Value to Round*

Use the concept web to write examples of rounding the number 284,761 to different place values.

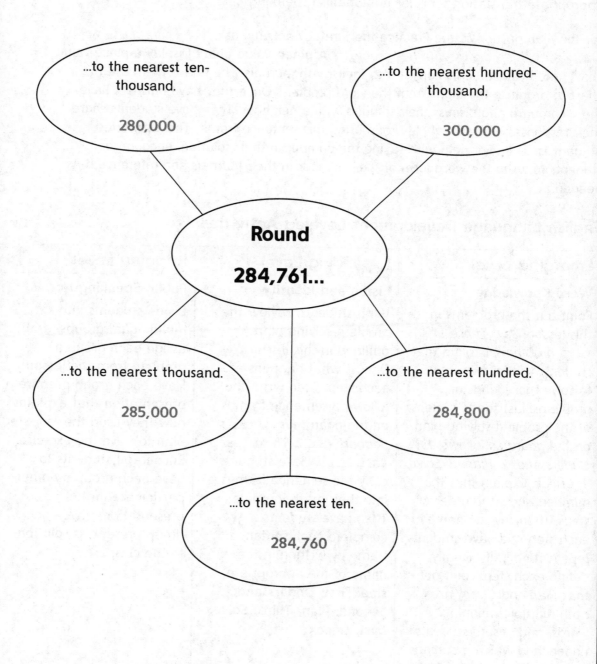

...to the nearest ten-thousand.

280,000

...to the nearest hundred-thousand.

300,000

**Round**

**284,761...**

...to the nearest thousand.

285,000

...to the nearest hundred.

284,800

...to the nearest ten.

284,760

# Lesson 6 Problem-Solving Investigation Strategy: Use the Four-Step Plan

## English Learner Instructional Strategy

### Collaborative Support: Turn and Talk

Before the lesson, write *problem* and its Spanish cognate, *problema*, on a cognate chart. Introduce the words and provide a concrete example. Utilize other appropriate translation tools for non-Spanish speaking ELs.

For the Plan portion of the Practice the Strategy activity, ask, *What graphic organizer could help us with this problem?* **A place-value chart** Display a place-value chart, and have students copy it into their math journals. Then have them use the chart to organize the digits from the word problem. Once they have finished, have them compare and discuss their solution with a neighbor. Then have students share their responses with the whole group using this sentence frame: **The estimated amount is _____.** For help with saying the number in their solution, encourage students to write the word form of their number in their journals and reference it as needed.

## English Language Development Leveled Activities

| Emerging Level | Expanding Level | Bridging Level |
|---|---|---|
| **Word Knowledge** | **Listen and Identify** | **Public Speaking Norms** |
| Point out the problems in this lesson. Say, *These are word problems.* Explain that students will follow a plan to solve these kinds of problems. List the four steps of the problem-solving plan on the board as follows: *1. Understand; 2. Plan; 3. Solve; 4. Check.* Explain that the numbers show the order of steps. Then say the name of each step and have students repeat it chorally. Finally, define each step. Point out that *Plan* and *Check* are both multiple-meaning words, and ensure students understand which meanings apply within a math context. | With students, review the problem-solving plan outlined in this lesson. Discuss what happens at each step. Then write the following sentence frames on the board: *First, _____; Second, _____; Third, _____; Last, _____.* Have students copy the sentence frames into their math journals. Then have students complete each sentence frame by writing in the name of the appropriate step: **First, Understand; Second, Plan; Third, Solve; Last, Check** | Divide students into multilingual groups, and assign each group a problem from the lesson. Have each group prepare a presentation that explains how they used the four-step plan to solve the exercise. Encourage students to have each group member participate in the presentation. Have each group present its solution to the class. |

**Teacher Notes:**

NAME _____ DATE _____

# Lesson 6 Problem-Solving Investigation
## STRATEGY: Use the Four-Step Plan

Solve each problem using the four-step plan.

| Prices of Cars | |
|---|---|
| Cars | Price |
| Car A | $83,532 |
| Car B | $24,375 |
| Car C | $24,053 |
| Car D | $73,295 |

1. **Mr. Kramer** is buying a car.
   The list of **prices** is shown in the table.
   Mr. Kramer wants to **buy** the **least** expensive car.
   Which car should **he** (Mr. Kramer) buy?

**Understand**

I know...

I need to find...

**Solve**

| Thousands | | | Ones | | |
|---|---|---|---|---|---|
| hundreds | tens | ones | hundreds | tens | ones |
| | | | , | | |
| | | | , | | |
| | | | , | | |
| | | | , | | |

**Plan**

I can use a ___place-value chart___
to help me solve.

**Check**

2. **A restaurant** made **more** than **$80,000** but **less** than **$90,000** **last** month.
   There is a **6** in the **ones** place, a **3** in the **thousands** place,
   a **7** in the **hundreds** place, and a **1** in the **tens** place.
   **How much money** did the restaurant make **last** month?

**Understand**

I know...

I need to find...

**Solve**

| Thousands Period | | | Ones Period | | |
|---|---|---|---|---|---|
| hundreds | tens | ones | hundreds | tens | ones |
| | | | , | | |

**Plan**

I can use a ___place-value chart___
to help me solve.

**Check**

Grade 4 • Chapter 1 *Place Value*  **7**

# Chapter 2 Add and Subtract Whole Numbers

## *What's the Math in This Chapter?*

**Mathematical Practice 8: Look for and express regularity in repeated reasoning.**

Prepare a demonstration place-value chart with the following numbers written: 536; 546; 556. Leave a black space below 556. Display the chart and say, *Look at the chart carefully. Do you see a pattern?*

Allow students time to think on their own. Then encourage them to turn and talk with a peer. Discuss as a group. The discussion goal is for students to identify the "add ten" pattern. Ask, *Now that we know the pattern is to add ten, what is the next number?* **566** Write 566 on the place-value chart.

Discuss with students the pattern of adding 10 is a repeated calculation. Say, *When you figured out that the next number was 566, you were using repeated reasoning.*

Display a chart with Mathematical Practice 8. Restate Mathematical Practice 8 and have students assist in rewriting it as an "I can" statement, for example: **I can solve problems by using repeated calculations.** Have students draw or write examples of using repeated reasoning/repeated calculations. Post examples and the new "I can" statement.

## *Inquiry of the Essential Question:*

**What strategies can I use to add or subtract?**

Inquiry Activity Target: **Students come to a conclusion that they can use repeated calculations to solve problems.**

As an introduction to the chapter, present the Essential Question to students. The inquiry graphic organizer will offer opportunities for students to observe, make inferences, and apply prior knowledge of repeated operations representing the Essential Question. As they investigate, encourage students to draw, write, and collaborate with peers to demonstrate their observations and thinking. Then have students present additional questions they may have to a peer to extend discussions.

Regroup students and restate Mathematical Practice 8 and the Essential Question. Pose questions to reflect on what has been learned to guide students in making connections between the Mathematical Practice and the Essential Question.

NAME _____ DATE _____

# Chapter 2 Add and Subtract Whole Numbers

## *Inquiry of the Essential Question:*

**What strategies can I use to add or subtract?**

Read the Essential Question. Describe your observations (I see...), inferences (I think...), and prior knowledge (I know...) of each math example. Write additional questions you have below. Then share your ideas and questions with a classmate.

$$
\begin{array}{r}
1{,}255 \\
+\,6{,}740
\end{array}
\;\boxed{\text{rounds to}}\;
\begin{array}{r}
1{,}000 \\
+\,7{,}000 \\
\hline
8{,}000
\end{array}
$$

I see ...

I think...

I know...

---

thousands  hundreds  tens  ones

$$
\begin{array}{r}
8\ \ 13 \\
5,\cancel{9}\ \cancel{3}\ 8 \\
-\quad 2\ 7\ 6 \\
\hline
5,\ 6\ 6\ 2
\end{array}
$$

I see ...

I think...

I know...

---

| 4,136 | 4,236 | 4,336 | |

+100   +100   +100

I see ...

I think...

I know...

---

Questions I have...

_____

_____

# Lesson 1 Addition Properties and Subtraction Rules

## English Learner Instructional Strategy

### Vocabulary Support: Frontload Academic Vocabulary

Before the lesson, have students preview the My Vocabulary Cards (also available in Spanish online) for *Commutative Property*, *Associative Property*, and *Identity Property*. Explain that *property* is a multiple-meaning word, and discuss how its meanings within the contexts of theater, real estate, and math both relate and differ. Use picture support.

To introduce vocabulary, write *commute/commutative* and *associate/associative* on the board. Say, *When we commute, we travel to and from. When we associate with others, we work with them.* Provide pictures to support meaning. Discuss how these meanings relate to the meanings of *Commutative Property* and *Associative Property*, respectively. Then point to the parentheses on the *Associative Property* card. Discuss differences in the way parentheses are used in these content areas: language arts/writing and math.

### English Language Development Leveled Activities

| Emerging Level | Expanding Level | Bridging Level |
|---|---|---|
| **Act It Out** Use connecting cubes to demonstrate these actions, and then have students mimic with their own connecting cubes: 1. *Start with 2 cubes and add 3 more; count to show that the total is 5; write 2 + 3 = 5.* 2. *Start with 3 cubes and add 2 more; count to show that the total is 5; write 3 + 2 = 5.* Point to the equations and say, *This shows the Commutative Property.* Have students chorally repeat, **Commutative Property**. Discuss how the order in which the numbers are added does not change the sum. | **Background Knowledge** On the board, write the following three pairs of equations: 4 + 8 = 12, 8 + 4 = 12; 3 + (2 + 9) = 14, (3 + 2) + 9 = 14; 7 + 0 = 7, 0 + 7 = 7 Point to each pair of equations, and have students identify the property of addition that is represented (**Commutative, Associative,** and **Identity**, respectively). Then write this subtraction sentence: 45 − ___ = 45. Ask students to find the unknown. **0** Repeat the exercise with additional equations and subtraction sentences with unknowns. | **Building Oral Language** Divide students into three multilingual groups, and assign each group one of the properties of addition. Then ask each group to discuss their assigned property and think of a strategy for remembering what the property states. Provide an example. Say, *The Commutative Property states that order doesn't matter. Commute means "move to and from," so it's easy to remember that the Commutative Property deals with moving numbers.* Have each group present its property and strategy to the class. |

**Teacher Notes:**

NAME _____   DATE _____

# Lesson 1 Vocabulary Chart
## *Addition Properties and Subtraction Rules*

Use the three-column chart to organize the vocabulary in this lesson.
Write the word in Spanish. Then write the correct terms to complete
each definition.

| English | Spanish | Definition |
|---|---|---|
| **Commutative Property of Addition** | propiedad conmutativa de la suma | The property that states that the order in which <u>two</u> numbers are added <u>does</u> <u>not</u> change the <u>sum</u>. <br> $12 + 15 = 15 + 12$ |
| **Associative Property of Addition** | propiedad asociativa de la suma | The property that states that the grouping of the addends <u>does</u> <u>not</u> change the <u>sum</u>. <br> $(4 + 5) + 2 = 4 + (5 + 2)$ |
| **Identity Property of Addition** | Propiedad de identidad de la suma | For any number, <u>zero</u> plus that number is the number. <br> $3 + 0 = \underline{3}$ or $0 + 3 = \underline{3}$ |
| **unknown** | incógnita | The amount that has not been <u>identified</u>. |

# Lesson 2 Addition & Subtraction Patterns
## English Learner Instructional Strategy

### Language Structure Support: Tiered Questions

For Math in My World, Example 1, utilize the following lower tiered questions: *How many people saw the new movie on Friday?* **1,323** *The word problem says 1,000 more people saw the movie on Saturday. Do we add or subtract 1,000?* **Add** *What do we get when we add 1,323 and 1,000?* **2,323** *The word problem says 100 less people saw the movie on Sunday. Do we add or subtract?* **Subtract** *Think about the number we subtract from. Do we subtract from Friday's number, 1,323?* **No** *Do we subtract from Saturday's number, 2,323?* **Yes** *How many people saw the movie on Sunday?* **2,223**

For Math in My World, Example 2, define *pattern* using realia. Show and discuss similarities and differences in how a pattern is defined within the content areas of art and math. Invite students to share their own examples of patterns seen in everyday life. **For example, The quilt on my bed has colors that are in a pattern.**

### English Language Development Leveled Activities

| Emerging Level | Expanding Level | Bridging Level |
|---|---|---|
| **Word Knowledge** Write this number on the board: 26,421. Say, *Let's find 100 more.* Then write this equation: *26,421 + 100 = 26,521.* Say, *To find the number that is 100 more, we add 100.* Circle the plus sign and say, *More.* Have students repeat chorally. Then say, *Let's find the number that is 1,000 less.* Write this equation: *26,421 − 1,000 = 25,421.* Say, *To find the number that is 1,000 less, we subtract 1,000.* Circle the minus sign and say, *Less.* Have students repeat chorally. Repeat the exercise, adding and subtracting powers of ten from other multi-digit numbers. | **Building Oral Language** Draw the following place-value chart:<br><br>Point to the number in the second row and ask, *How many more?* Guide students to say, **10,000.** Replace that number with *75,342.* Ask, *How many less?* Guide students to respond, **100 less.** Use other numbers to repeat. | **Number Game** Have each student write a six-digit number and a series of five addition/ subtraction instructions using powers of ten. For example: **10,000 more; 1,000 more; 10 less; 100 more; 1,000 less.** Have students exchange papers with a partner and follow each other's instructions to find the resulting number. Have partners check each other's work. |

Place-value chart (Expanding Level):

| Thousands | | Ones | | |
|---|---|---|---|---|
| tens | ones | hundreds | tens | ones |
| 7 | 5 | 4 | 4 | 2 |
| 8 | 5 | 4 | 4 | 2 |

### Teacher Notes:

NAME _____ DATE _____

# Lesson 2 Note Taking
## *Addition and Subtraction Patterns*

Read the question. Write words you need help with and research each word. Use your lesson to write your Cornell notes. Write or draw math examples to explain your thinking. Share your examples with a classmate.

**Building on the Essential Question**

How can patterns help with addition or subtraction?

**Words I need help with:**

See students' words.

**My Math Examples:**

See students' examples.

**Notes**

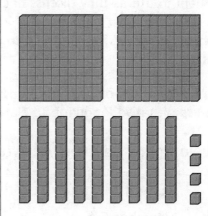

Complete the place value chart for the amount of blocks shown.

| hundreds | tens | ones |
|----------|------|------|
| 2 | 9 | 4 |

To find *100 more* than the amount of blocks shown, you use the operation of __addition__.

100 more than the amount of blocks shown is __394__.

Another 100 more is equal to __494__.

To find *10 less* than the amount of blocks shown, you use the operation __subtraction__.

10 less than the amount of blocks shown is __284__.

Another 10 less is equal to __274__.

# Lesson 3 Add and Subtract Mentally
## English Learner Instructional Strategy

### Language Structure Support: Communication Guide

Before having students answer the Talk Math question, point out that *difference* is a multiple-meaning word that can mean "the quality of being not the same" or "an argument." Then say, *In math, the meaning of difference is "the number that remains after subtraction."*

Assign partners to work on Independent Practice Exercises 6–9. Provide this communication guide for them to use as they discuss each problem: **To solve this problem, we can make a (ten /hundred /thousand). First, add _____ to _____ to make _____. Then subtract _____ from _____ to make _____. Add _____ to _____. The solution is _____.**

Have students work with a bilingual aid or peer to discuss and solve Problem Solving Exercises 18–21. Encourage use of native language to clarify meaning.

### English Language Development Leveled Activities

| Emerging Level | Expanding Level | Bridging Level |
|---|---|---|
| **Word Knowledge** Solve this problem, writing each step of the solution: 567 + 354. Then say, *Sometimes we solve equations by writing.* Move your hand to indicate writing, and say, *Solve by writing.* Have students repeat chorally. Then think aloud to solve this problem: 421 − 360. Say, *Round 421 to 420, subtract 360 to get 60, and then add 1 to get the answer, 61.* Then say, *Sometimes we don't need to write to solve a problem. Instead, we can solve it mentally, or think of the answer.* Tap the side of your head, and say, *Solve mentally.* Have students repeat chorally. | **Number Sense** Write this problem on the board: 82 − 6. Ask, *How can we make 82 into a ten?* Guide students to respond, **82 rounds to 80 to make a ten.** Think aloud to demonstrate solving the problem mentally from there. Then use the following problems to have students describe making a hundred and a thousand. 693 + 28: **Make a hundred by rounding 693 to 700.** 4,039 − 550: **Make a thousand by rounding 4,039 to 4,000.** Encourage students to "think aloud" to demonstrate solving both problems mentally. | **Public Speaking Norms** Divide students into three multilingual groups, and assign each group a different word problem from Problem Solving Exercises 18–20. Then ask each group to solve their assigned problem and present their solution to the group. Remind students to explain how they used mental math. |

### Teacher Notes:

NAME _____ DATE _____

# Lesson 3 Guided Writing

## *Add and Subtract Mentally*

**How do you add and subtract mentally?**

Use the exercises below to help you build on answering the Essential Question. Write the correct word or phrase on the lines provided.

1. Rewrite the question in your own words.
   <u>See students' work.</u>
   _____

2. What key words do you see in the question?
   <u>add, subtract, mentally</u>

3. If you add <u>3</u> to 147, you make 150, which ends in a <u>ten</u> .

4. If you add 3 to one addend, what must you do to the other addend?
   <u>subtract 3</u>

5. What problem will help you mentally solve 147 + 26? <u>150 + 23 = 173</u>

6. If you subtract <u>12</u> from 312, you make 300, which ends in a <u>hundred</u> .

7. What problem will help you mentally solve 312 − 241? <u>300 − 241 = 59</u>

8. When you subtract 12 from 312 to help you mentally solve 312 − 241, you must <u>add</u> 12 to the difference. 300 − 241 = <u>59</u> + 12 = <u>71</u>

9. Mentally solve 441 − 302. <u>139</u>

10. How do you add and subtract mentally?
    <u>You can take and give to make one number end in a ten, hundred,</u>
    <u>or thousand.</u>

# Lesson 4 Estimate Sums and Differences
## English Learner Instructional Strategy

### Vocabulary Support: Anchor Chart

Before the lesson, write the following words on a cognate chart, along with their Spanish cognates: *estimate/estimar, sum/suma, difference/diferencia*. Introduce the words, and provide realia to support understanding. Utilize other appropriate translation tools for non-Spanish speaking ELs.

Then have students help you create an Anchor Chart for *sum* and *difference*. Ask volunteers for two equations to write, one that uses addition and one that uses subtraction. Circle and label the sum in the addition equation and the difference in the subtraction equation.

Finally, review the definition for *round*. Then guide students to round in order to estimate the answer to each equation on the Anchor Chart. Write both estimates on the chart in boxes labeled *Estimate*.

### English Language Development Leveled Activities

| Emerging Level | Expanding Level | Bridging Level |
|---|---|---|
| **Developing Oral Language** Write: 5,481 + 2,326. Demonstrate rounding to the nearest hundred (5,500 + 2,300) to find an estimated sum. Then say, *5,481 + 2,326 is about 7,800.* Emphasize the word *about.* Now write: 5,481 + 2,326 = 7,807. Say, *5,481 + 2,326 is exactly 7,807.* Emphasize the word exactly. Finally, point to the estimated sum and say, *This is an* **estimate**. Have students repeat chorally. Point to the actual sum and say, *This is an* **exact** *answer.* Have students repeat chorally. | **Listen and Identify** Explain that an estimate tells *about* how much, while an exact answer tells *exactly* how much. Then write this on the board: 63,718 + 19,007. Guide students in rounding each addend to the nearest ten thousand. Write: 60,000 + 20,000 = 80,000. Then say, *63,718 + 19,007 is about 80,000.* Emphasize the word *about.* Now guide students in rounding each addend to the nearest thousand. Say this new estimated sum: *64,000 + 19,000 is about 83,000.* Repeat these exercises with other multi-digit addition problems. | **Show What You Know** Write this problem: 42,036 − 11,889. Then divide students into four groups. Have each group estimate the solution by rounding to an assigned place value; assign one group ten thousands, another group thousands, the third group hundreds, and the last group tens. Have each group present its estimate to the class using the following sentence frame: **42,036 − 11,889 is about ____. We rounded to the nearest ____.** |

**Teacher Notes:**

NAME _____ DATE _____

# Lesson 4 Vocabulary Definition Map
## *Estimate Sums and Differences*

Use the definition map to write a description and list characteristics about the vocabulary word or phrase. Write or draw math examples. Share your examples with a classmate.

My Math Vocabulary:

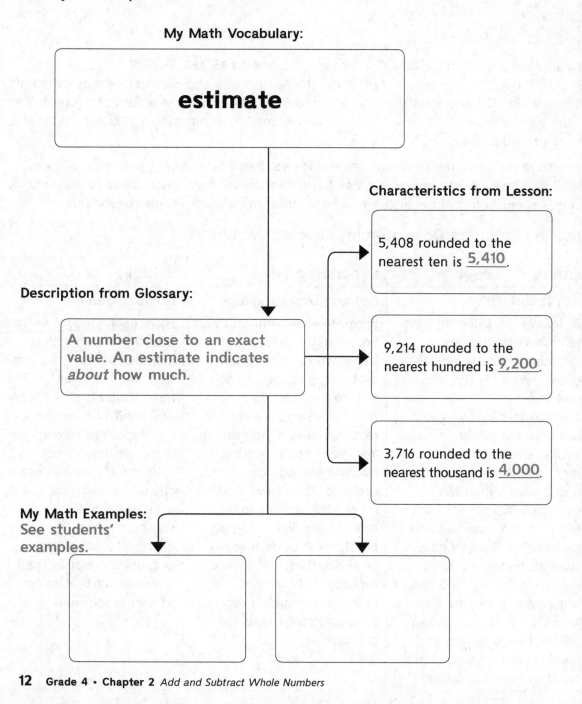

**estimate**

**Characteristics from Lesson:**

5,408 rounded to the nearest ten is <u>5,410</u>.

9,214 rounded to the nearest hundred is <u>9,200</u>.

3,716 rounded to the nearest thousand is <u>4,000</u>.

**Description from Glossary:**

A number close to an exact value. An estimate indicates *about* how much.

**My Math Examples:**
See students' examples.

# Lesson 5 Add Whole Numbers

## English Learner Instructional Strategy

### Vocabulary Support: Frontload Academic Vocabulary

Before the lesson, create the following table.

| Estimate | Answer |
|----------|--------|
| 15,000 | 15,236 |
| 110,000 | 8,030 |
| 2,000 | 6,213 |

Discuss how the estimate 15,000 is close to the answer 15,236. Say, *15,000 is close to 15,236. The estimate is reasonable.* Stress the terms *close* and *reasonable*. Have students compare 110,000 and 8,030, and 2,000 and 6,213 respectively. Ask for both comparisons, *Is the estimate close to the actual answer?* **No** Say, *Then the estimate is not reasonable.* Stress the word *not*.

Have students write the following sentences into their Math Journals then read them aloud to a peer. **If my answer is close to the estimate, then my answer is reasonable. If my answer is not close to the estimate, then my answer is not reasonable.**

### English Language Development Leveled Activities

| Emerging Level | Expanding Level | Bridging Level |
|----------------|-----------------|----------------|
| **Act It Out** | **Building Oral Language** | **Number Game** |
| Distribute the following base-ten blocks to students: 1 hundreds, 11 tens, and 13 ones. Then write this problem on the board: 53 + 55. Have students mimic you as you use base-ten blocks to model adding the ones and regrouping 13 ones as 1 ten and 3 ones. When you exchange 10 ones for 1 ten, say, *regroup.* Have students repeat chorally. Now have students mimic you as you model adding the tens and regrouping 11 tens as 1 hundred and 1 ten. When you exchange 10 tens for 1 hundred, say, *regroup.* Have students repeat chorally. | Write this problem on the board: 629 + 543. Demonstrate solving the problem, pausing at each step to ask, *Should I regroup?* Guide students to respond correctly by saying, **Yes, you should regroup,** or **No, you should not regroup.** Then have students repeat the activity with other multi-digit addition problems. Provide these sentence frames: **I should regroup. I should not regroup. I shouldn't regroup.** Encourage the use of the contraction *shouldn't.* | Have each student write a five-digit number that includes at least two digits that are 6 or greater. Then have students pair up and add their two numbers, regrouping as needed to arrive at the correct solution. Have pairs use a calculator to check solutions. Finally, have students repeat the exercise, adding their original number to that of a different partner to arrive at a new solution. |

NAME _____ DATE _____

# Lesson 5 Concept Web
## *Add Whole Numbers*

Use the concept web to identify parts of finding the sum of whole numbers. Use the terms from the word bank.

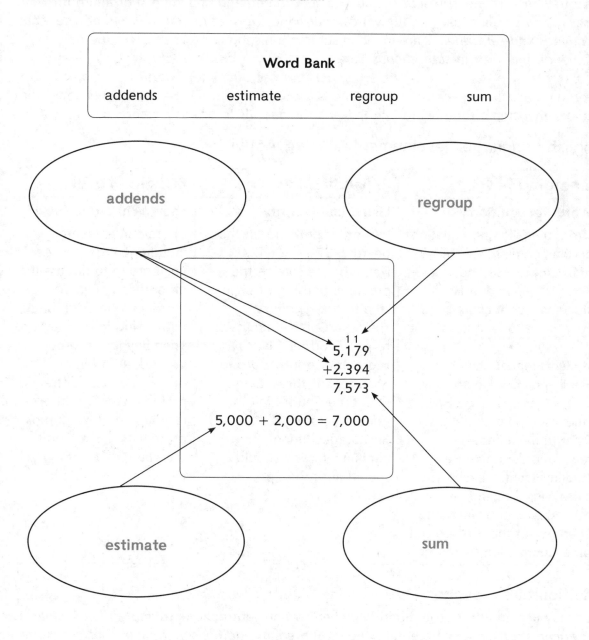

**Word Bank**

addends      estimate      regroup      sum

addends

regroup

$$\begin{array}{r} 1\ 1 \\ 5{,}179 \\ +2{,}394 \\ \hline 7{,}573 \end{array}$$

$$5{,}000 + 2{,}000 = 7{,}000$$

estimate

sum

Grade 4 • **Chapter 2** *Add and Subtract Whole Numbers*   **13**

# Lesson 6 Subtract Whole Numbers
## English Learner Instructional Strategy

### Sensory Support: Mnemonics

Before the lesson, write *minuend* and *subtrahend* on the board. Define both words using My Vocabulary Cards. Then give students the following mnemonic devices to help them remember how the words are defined. Write *minus* next to *minuend* and underline *minu* in both words. Then write *subtract* next to *subtrahend* and underline *subtra* in both words. Discuss how students can use the words *minus* and *subtract* to help them remember that *minuends* and *subtrahends* are found in subtraction sentences. Then circle the *m* in *minuends* and the *s* in *subtrahends*. Say, *If you can remember that m comes before s in the English alphabet, it will help you remember that the minuend is the first number in a subtraction sentence and the subtrahend is the second number.* Have students copy what you have written on the board into their Math Journals to use as a reference.

### English Language Development Leveled Activities

| Emerging Level | Expanding Level | Bridging Level |
|---|---|---|
| **Word Recognition** | **Listen and Identify** | **Show What You Know** |
| Write the following equation on the board: $220 - 74 = 146$. Point to 220 and say, *This is the* **minuend**. Have students repeat chorally. Then point to 74 and say, *This is the* **subtrahend**. Have students repeat chorally. Finally point to *146* and say, *This is the* **difference**. Have students repeat chorally. Point to the numbers in random order and have students identify each by saying, **minuend, subtrahend,** or **difference**. Then repeat the exercise with other subtraction sentences. | Write this problem on the board: $5,394 - 2,647$. Demonstrate solving the problem, pausing at each step to ask, *Should I regroup?* Guide students to respond correctly by saying, **regroup** or **don't regroup.** Then repeat the activity with other multi-digit addition problems. Encourage students to explain why you should or should not regroup. | Divide students into multilingual groups. Ask each group to discuss the difference between regrouping in addition and regrouping in subtraction. Then have each group prepare an addition problem and a subtraction problem where both require regrouping. Have groups present their problems to the class, explaining each instance of regrouping. |

### Multicultural Teacher Tip

ELs may use an alternative algorithm when solving subtraction problems. In particular, Latin American students may have been taught the *equal additions method* of subtraction instead of the traditional US method of "borrowing" from the column to the left when the top number is less than the bottom number.

In the *equal additions method*, a problem such as $35 - 18$ solved vertically would start with ten ones added to the top number ($15 - 8$) and then one ten is added to the bottom number ($30 - 20$), to get 7 and 10, or 17. Similarly, $432 - 158$ would be solved as $12 - 8$, $130 - 60$, and $400 - 200$ ($4 + 70 + 200 = 274$).

NAME _____ DATE _____

# Lesson 6 Vocabulary Cognates
## *Subtract Whole Numbers*

Use the Glossary to define the math word in English and in Spanish in the word boxes. Write a sentence using your math word.

| minuend | minuendo |
|---|---|
| **Definition**<br>The first number in a *subtraction* sentence from which a second number is to be subtracted. | **Definición**<br>El primer número en un enunciado de *resta*, del que se restará un segundo número. |

**My math word sentence:**

Sample answer: In the subtraction sentence 8 − 3 = 5, the minuend is 8.

| subtrahend | sustraendo |
|---|---|
| **Definition**<br>A number that is *subtracted* from another number. | **Definición**<br>Un número que se *resta* de otro número. |

**My math word sentence:**

Sample answer: In the subtraction sentence 14 − 5 = 9, the subtrahend is 5.

# Lesson 7 Subtract Across Zeros
## English Learner Instructional Strategy

### Sensory Support: Manipulatives

Write *zero* and its Spanish cognate, *cero* on a cognate chart. Introduce the word, and provide realia to model an instance of having zero to support understanding. Utilize other appropriate translation tools for non-Spanish speaking ELs.

Before the lesson, have a bilingual peer or mentor use modeled talk and base-ten blocks to review regrouping. Have them model: 1 ten = 10 ones, 1 hundred = 10 tens, and 1 thousand = 10 hundreds. Then have the EL remodel the process back to the peer/mentor.

Have ELs use base-ten blocks to help model while verbalizing their answer to the Talk Math prompt.

### English Language Development Leveled Activities

| Emerging Level | Expanding Level | Bridging Level |
|---|---|---|
| **Phonemic Awareness** | **Listen and Identify** | **Number Game** |
| On the board, write *0* and say, *One zero.* Have students repeat chorally. Write another *0* and say, *Two zeros.* Emphasize the /z/ at the end of *zeros.* Have students repeat chorally, and listen for the correct ending sound. Then write a variety of multi-digit numbers that contain multiple zeros. With students, count the number of zeros in each number, and then state the total. For example: Write *300,000*; count with students, *One, two, three, four, five zeros;* and then repeat, *Five zeros.* Have students repeat chorally. | Distribute base-ten blocks to each student. Then write the following problem on the board: 300 − 78. Guide students in using their base-ten blocks to regroup 3 hundreds into 2 hundreds and 10 tens. Then guide them to regroup 10 tens into 9 tens and 10 ones. Finally, have students take away 8 ones and 7 tens from their set of base-ten blocks to find the solution to the problem, 222. | On the board, write a subtraction sentence that requires subtracting across zeros (for example: 500 − 96). Say, *The first student to give the answer and correctly explain how to regroup gets to write the next subtraction problem.* Have students solve the problem independently. Then repeat the activity as needed for practice. |

**Teacher Notes:**

NAME _____  DATE _____

# Lesson 7 Note Taking
## *Subtract Across Zeros*

**Read the question. Write words you need help with and research each word. Use your lesson to write your Cornell notes. Write or draw math examples to explain your thinking. Share your examples with a classmate.**

| Building on the Essential Question | Notes: |
|---|---|
| How do you subtract across zeros? | When you subtract from 2,015, start in the <u>ones</u> place. Then subtract the <u>tens</u> place. Then subtract the <u>hundreds</u> place, and finally subtract the <u>thousands</u> place.<br><br>You can regroup 1 ten as 10 <u>ones</u>.<br><br>You can regroup 1 hundred as 10 <u>tens</u>.<br><br>You can regroup 1 thousand as 10 <u>hundreds</u>.<br><br>Complete the equation below to find the difference. |

**Words I need help with:**

See students' words.

$$\begin{array}{r} \boxed{1}\ \boxed{10}\phantom{00} \\ 2,\cancel{0}\ 1\ 5 \\ -\ 1,\ 3\ 1\ 4 \\ \hline \boxed{\phantom{0}},\boxed{7}\ \boxed{0}\ \boxed{1} \end{array}$$

**My Math Examples:**

See students' examples.

# Lesson 8 Problem-Solving Investigation
# Strategy: Draw a Diagram
## English Learner Instructional Strategy

### Sensory Support: Manipulatives

Before the lesson, write this addition sentence on the board: *10 + 4 = 14.* Use connecting cubes to model adding 10 cubes to 4 cubes to make a chain of 14 cubes. Say, *This whole chain equals 14 cubes.* Then pull apart the cubes again, so that you have a chain of 10 cubes and a chain of 4 cubes. Say, *I pulled the chain into two parts. One part has 10 cubes, and the other part has 4 cubes.* Then draw a bar diagram on the board, minus the line that divides it into parts. Say, *This is a bar diagram. We can use it to show the same equation I modeled with the connecting cubes.* Above the bar write 14, and say, *This whole bar equals 14.* Then draw a line to divide the bar into two parts, writing 10 in the first part and 4 in the second part. Point to each part and say, *Two parts make up the whole bar. One part equals 10 and the other part equals 4.* Show how to use a bar diagram to illustrate equations, such as these, that include an unknown whole or part:

_____ + 2 = 5; 3 + 2 = _____; 6 − 2 = _____; 6 − _____ = 4. Encourage student input.

### English Language Development Leveled Activities

| Emerging Level | Expanding Level | Bridging Level |
|---|---|---|
| **Word Knowledge** | **Background Knowledge** | **Show What You Know** |
| <br>&#124;---------- 7,894 ----------&#124;<br>&#124; 2,631 &#124; 5,263 &#124;<br><br>Draw the above bar diagram. Say, *This is a bar diagram.* Point to the 7,894 then write *whole* above the number. Have students chorally say, **whole**. Then point at 2,631 and 5,263 and say, *two parts.* Label both numbers *part.* Write the numbers in an addition sentence. Point to each number and have the students identify it as a *part* or a *whole.* | <br>&#124;---------- Whole ----------&#124;<br>&#124; Part &#124; Part &#124;<br><br>Draw the above bar diagram. Write, then say aloud: *There are 243 trees in Grant Park. 75 of them are pine trees. How many are not pine trees?* Guide students in using the bar diagram to solve the problem. Help them to recognize that the whole is 243, one part is 75, and the other part is found by subtracting, as follows: 243 − 75 = 168. Repeat the exercise with other simple word problems. | Pair each student with an emerging/expanding level student. Assign Exercise 1 from the lesson. Have bridging level students read the problem aloud and then draw a bar diagram to solve their assigned problems. Then have bridging level students assist their partner with Exercise 2. |

### Teacher Notes:

NAME _____ DATE _____

# Lesson 8 Problem-Solving Investigation
## STRATEGY: Draw a Diagram

Draw a diagram to help you solve the problems.

1. **Twickenham Stadium,** in England, can seat **82,000** people.
   If there are **49,837** people seated in the stadium,
   how many **more people** can be seated in the **stadium?**

| Understand | Solve |
|---|---|
| I know:<br><br>I need to find: | |
| **Plan**<br><br>├------------ 82,000 ------------┤<br>[ 49,837 \| ? ] | **Check** |

2. A bakery uses **ten cups of butter** and ten **eggs** for a recipe.
   There are **16,280 Calories** in **ten cups of butter.**
   **Ten eggs** have **1,170 Calories.**
   **How many more** Calories are there in ten cups of butter **than** in ten eggs?

| Understand | Solve |
|---|---|
| I know:<br><br>I need to find: | |
| **Plan**<br><br>├------------ 16,280 ------------┤<br>[ 1,170 \| ? ] | **Check** |

# Lesson 9 Solve Multi-Step Word Problems
## English Learner Instructional Strategy

### Vocabulary Support: Modeled Talk

Write *equation* and *variable* on a cognate chart, along with their Spanish cognates, *ecuación* and *variable*, respectively. Introduce the words, and provide visual examples to support understanding. Utilize other appropriate translation tools for non-Spanish speaking ELs.

Write this word on the board: *multi-step*. Underline *multi-* then discuss using Modeled Talk: *Multi- is a word part that means "more than one."* Then underline *step* and say, *Step is a multiple-meaning word. Here it means "one part or movement in a process." Together multi- and step create a word used to describe a process that has more than one step, or part.* Show examples of multi-step processes, such as recipes and instructions for putting together toys. Then invite students to say and define other words they know that begin with *multi-*, such as **multi-grain** and **multi-colored**. Also, discuss how the meaning of step differs in other content areas, such as dance and architecture.

### English Language Development Leveled Activities

| Emerging Level | Expanding Level | Bridging Level |
|---|---|---|
| **Word Knowledge** | **Act It Out** | **Synthesis** |
| Write the following equation: $26 - 15 = 11$. Point to the (=) equals symbol, and say, *equals.* Then gesture across the entire equation, and say, *equation.* Have students repeat chorally. Now write this on the board: $7 + 16$. Say, *There is no equals sign, so this is not an equation.* Have students repeat chorally, **not an equation.** Write other examples and non-examples of equations on the board. Have students distinguish between them by saying, **equation** or **not an equation.** | Distribute a handful of counters to each student. Have each student find the total number of counters he or she has (for example, **17**) and record that number. Then have students divide their counters into two piles. Have them count what is in Pile 1 and record the number (for example, **4**). Then tell students, *The number of counters in Pile 2 is unknown.* Explain how you can discover the number of counters in Pile 2 using an equation that includes a variable to represent the unknown (for example, $4 + x = 17$). | Have multilingual groups work together to create a word problem, referring to the lesson for examples. Then ask them to exchange word problems with another group. Tell each group to solve the word problem they received using an equation that includes a variable for the unknown. Then have groups present their equations and solutions to the class. |

**Teacher Notes:**

NAME _____ DATE _____

# Lesson 9 Four-Square Vocabulary
## *Solve Multi-Step Word Problems*

Write the definition for each math word. Write what each word means in your own words. Draw or write examples that show each math word meaning. Then write your own sentences using the words.

| **Definition** | **My Own Words** |
|---|---|
| A sentence that contains an equals sign (=), showing that two *expressions* are equal. | See students' examples. |

equation

| **My Examples** | **My Sentence** |
|---|---|
| Sample answer: The equation $3 + 1 = 4$ shows that adding one more to three is equal to four. | Sample sentence: An equation can be used to solve addition and subtraction problems. |

| **Definition** | **My Own Words** |
|---|---|
| A letter or symbol used to represent an unknown quantity. | See students' examples. |

variable

| **My Examples** | **My Sentence** |
|---|---|
| Sample answer: In the equation $3 + x = 4$, the variable $x$ represents 1. | Sample sentence: If I want to know how much more I need to reach 20, I can use a variable, $15 + x = 20$. |

**Grade 4 • Chapter 2** *Add and Subtract Whole Numbers* **17**

# Chapter 3 Understand Multiplication and Division

## What's the Math in This Chapter?

**Mathematical Practice 2: Reason abstractly and quantitatively.**

Write the numbers 1, 3, and 5 along with two multiplication signs on five pieces of paper. Have 5 students come to the front of the room and distribute a paper to each of them. Say, Create a multiplication sentence. Allow students time to place themselves in any order.

Say, *We are going to solve this multiplication problem.* Have students chorally solve the problem. For example, **One times three is three. Three times five is fifteen.** Write the multiplication sentence on the board. Have students rotate positions and repeat the process until all possible number sentences are represented.

Ask students to look carefully at the written multiplication sentences. Discuss their observations. The discussion goal is for students to discover that no matter what order the factors are multiplied in, the product is always the same. Explain to students that the Commutative Property of Multiplication says that changing the order when you multiply doesn't affect the product.

Display a chart with Mathematical Practice 2. Restate Mathematical Practice 2 and have students assist in rewriting it as an "I can" statement, for example: **I can use properties to better understand problems.** Post the new "I can" statement in the classroom.

## Inquiry of the Essential Question:

### How are multiplication and division related?

Inquiry Activity Target: **Students come to a conclusion that using properties can help them solve multiplication and division problems more easily.**

As an introduction to the chapter, present the Essential Question to students. The inquiry graphic organizer will offer opportunities for students to observe, make inferences, and apply prior knowledge of properties of operations representing the Essential Question. As they investigate, encourage students to draw, write, and collaborate with peers to demonstrate their observations and thinking. Then have students present additional questions they may have to a peer to extend discussions.

Regroup students and restate Mathematical Practice 2 and the Essential Question. Pose questions to reflect on what has been learned to guide students in making connections between the Mathematical Practice and the Essential Question.

NAME _____ DATE _____

# Chapter 3 Understand Multiplication and Division

## Inquiry of the Essential Question:

**How are multiplication and division related?**

Read the Essential Question. Describe your observations (I see...), inferences (I think...), and prior knowledge (I know...) of each math example. Write additional questions you have below. Then share your ideas and questions with a classmate.

| | | | |
|---|---|---|---|
| $2 \times 6 = 12$ | | $12 \div 2 = 6$ | |
| $6 \times 2 = 12$ | | $12 \div 6 = 2$ | |

I see …

I think…

I know…

---

**Words:**   5 times more than $4

|-------- $20 --------|

**Diagram:**

| 4 | 4 | 4 | 4 | 4 |
|---|---|---|---|---|

**Equation:** $5 \times \$4 = \$20$

I see …

I think…

I know…

---

$$
\begin{array}{ccccc}
30 & 24 & 18 & 12 & 6 \\
-\ 6 & -\ 6 & -\ 6 & -\ 6 & -6 \\
\hline
24 & 18 & 12 & 6 & 0
\end{array}
$$

I see …

I think…

I know…

---

Questions I have…

_____

_____

_____

# Lesson 1 Relate Multiplication and Division
## English Learner Instructional Strategy

### Vocabulary Support: Graphic Organizer

Write *division* and *multiplication* and their Spanish cognates, *divisiones* and *multiplicación*, respectively, on a cognate chart. Introduce the words, and provide a visual math example to support understanding. Utilize other translation tools for non-Spanish speaking ELs. Discuss these related word pairs: *division/divide, multiplication/multiply* Create a two-column chart with these heads: Column 1-*Multiplication*, Column 2-*Division*. Have students preview the My Vocabulary Cards for *dividend, divisor, quotient, product, factor,* and *fact family.* Then say, *Let's organize these words. Which words relate to multiplication?* **fact family, factor, product** Write each in Column 1. Say, *Which words relate to division?* **dividend, divisor, quotient, fact family** Write each in Column 2. Discuss how *fact family* appears in both columns. Post for reference.

### English Language Development Leveled Activities

| Emerging Level | Expanding Level | Bridging Level |
|---|---|---|
| **Word Knowledge** | **Developing Oral Language** | **Show What You Know** |
| Write these numbers on the board: 3, 4, 12. Then beneath the numbers, write this equation: $3 \times 4 = 12$. Point to the multiplication symbol and say, *This means multiplication.* Have students repeat chorally. Repeat the previous step using this equation: $4 \times 3 = 12$. Then write this equation: $12 \div 4 = 3$. Point to the division symbol and say, *This means division.* Have students repeat chorally. Repeat the previous step using this equation: $12 \div 3 = 4$. Finally, circle all four equations and say, *These are a fact family.* Have students repeat chorally. | Write the following numbers on the board: 6, 7, 42. Say, *I want to write the fact family for these three numbers.* Explain that a fact family is a set of facts that use the same three numbers. Guide students in identifying that $6 \times 7 = 42$ is one multiplication fact in the fact family. Have students use the following sentence frame to help them say the remaining facts: **A multiplication/division fact is ___.** Repeat the exercise with other sets of three numbers, such as 3, 9, 27. | Give each student a number cube. Tell students to roll the number cube twice and use each number they roll as a factor in a multiplication sentence, such as $2 \times 6 = 12$. Then have students write the remaining sentences in the fact family. (For example: $6 \times 2 = 12$; $12 \div 6 = 2$; $12 \div 2 = 6$) Ask each group to present its fact family to the class. |

### Multicultural Teacher Tip

Word problems are an important part of the math curriculum, but they can be particularly challenging for ELs, and not just because of language issues. Allow students to share examples from their own cultures, including popular national sports or physical activities they participated in, foods and drinks from their culture, traditional clothing worn in their home countries, and so on. When appropriate, help ELs reword an exercise to include a familiar cultural reference.

NAME _____ DATE _____

# Lesson 1 Vocabulary Chart
## *Relate Multiplication and Division*

Use the three-column chart to organize the vocabulary in this lesson.
Write the word in Spanish. Then write the correct terms to complete
each definition.

| English | Spanish | Definition |
|---|---|---|
| **dividend** | dividendo | A number that is being <u>divided</u>. |
| **divisor** | divisor | The number by which the <u>dividend</u> is being <u>divided</u>. |
| **factor** | factor | A number that <u>divides</u> a whole number <u>evenly</u>. Also a number that is <u>multiplied</u> by another number. |
| **product** | producto | The answer or result of a <u>multiplication</u> problem. It also refers to expressing a number as the <u>product</u> of its factors. |
| **quotient** | cociente | The result of a <u>division</u> problem. |
| **fact family** | familia de operaciones | A group of <u>related</u> facts using the same <u>numbers</u>. |

# Lesson 2 Relate Division and Subtraction

## English Learner Instructional Strategy

### Vocabulary Support: Make Connections

Before the lesson, draw a K-W-L Chart on chart paper. Then write the following: $3 + 3 + 3 = 9$. Say, *This equation includes repeated addition. What is repeated addition?* **Adding the same number again and again** Write *repeated addition*, along with students' definitions and your example equation, in the K column of the chart. Then say, *Multiplication is repeated addition.* Point to the equation you wrote. Ask, *How can we write this as a multiplication problem?* **3 × 3 = 9** Add *multiplication* and your example multiplication sentence to the K column of the chart. Then write the following in the W column of the chart: *What is repeated subtraction? How is repeated subtraction used in division?* Guide students to define *repeated subtraction*, using their definition for *repeated addition*: **Subtracting the same number again and again** Then say, *In this lesson we will learn how division uses repeated subtraction.* Have students help you complete the L column of the chart following the lesson.

### English Language Development Leveled Activities

| Emerging Level | Expanding Level | Bridging Level |
|---|---|---|
| **Word Knowledge** | **Act It Out** | **Public Speaking Norms** |
| Make a simple motion, such as sitting and then standing. Then say, *I will repeat*, and make the motion again. Do this several times, and then explain that *repeat* means "to do again." Now write 15 ÷ 3 on the board. Use repeated subtraction to find the answer to this problem. First write 15 − 3 vertically, and write the difference, 12. Then say, *I will repeat, to subtract 3 from 12.* Have students repeat chorally. Use the established routine to continue subtracting 3 until the difference is 0. | Divide students into three groups, and distribute 36 counters to each group. Then guide students in using the counters to find 36 ÷ 3 by repeatedly subtracting 3. For each subtraction of 3, write the corresponding equation as you say, *36 − 3 = 33, 33 − 3 = 30,* and so on. After each subtraction of 3, have groups make a new pile of 3 counters. Continue until each group has 12 piles. Then assign each group one of these problems to complete on their own: 36 ÷ 2; 36 ÷ 4; 36 ÷ 6. Have group members use repeated subtraction to find the solutions. | Have multilingual groups of students create a presentation that illustrates using repeated subtraction to divide. Assign a division problem, or allow groups to write their own problem. Then discuss how students might use a number line or manipulatives to model using repeated subtraction. Have each group present their problem using modeling and written calculations to show the solution to the class. |

### Teacher Notes:

NAME _____ DATE _____

# Lesson 2 Guided Writing

## *Relate Division and Subtraction*

**How do you relate division and subtraction?**

Use the exercises below to help you build on answering the Essential
Question. Write the correct word or phrase on the lines provided.

1. Rewrite the question in your own words.
   <u>See students' work.</u>
   _____

2. What key words do you see in the question?
   <u>subtraction, division, relate</u>

3. Repeated addition is <u>adding</u> the <u>same</u> number again and again.

4. The operation of <u>multiplication</u> is the same as repeated <u>addition</u>.

5. Repeated subtraction is <u>subtracting</u> the <u>same</u> number again and again.

6. How many times can you subtract 4 from 12 before reaching 0? Use the number
   line to help you.
   <u>3 times</u>

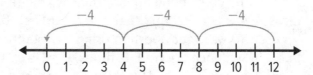

7. What is 12 ÷ 4?
   <u>3</u>

8. How do you relate division and subtraction?
   <u>Sample answer: I can use repeated subtraction to divide.</u>
   _____

**20** **Grade 4 · Chapter 3** *Understand Multiplication and Division*

# Lesson 3 Multiplication as Comparison
## English Learner Instructional Strategy

### Language Structure Support: Sentence Frames

Before the lesson, write then say: $2 \times 9 = 18$. Ask, *What is the first factor in this equation?* **2** Label the 2 *Factor 1.* Then ask, *What is the second factor in this equation?* **9** Label the 9 *Factor 2.* Finally ask, *What is the product in this equation?* **18** Label the 18 *Product.*

Then say, *We can use words to say multiplication sentences in several different ways.* Display these sentence frames on a chart:

1. <u>Factor 1</u> times <u>Factor 2</u> equals <u>Product</u>.
2. <u>Factor 1</u> times as many as <u>Factor 2</u> is <u>Product</u>.
3. <u>Factor 1</u> times more than <u>Factor 2</u> is <u>Product</u>.
4. <u>Factor 1</u> times as much as <u>Factor 2</u> is <u>Product</u>.

Have students practice applying the multiplication sentence, $2 \times 9 = 18$ as you point to each sentence frame. Encourage them to add the chart to their Math Journals.

### English Language Development Leveled Activities

| Emerging Level | Expanding Level | Bridging Level |
|---|---|---|
| **Word Recognition** | **Developing Oral Language** | **Number Game** |
| Prepare four sheets of paper, each depicting one of the following: a digital clock; an analog clock; the multiplication sentence, $3 \times 4 = 12$; and an array that represents the multiplication sentence, $2 \times 5$. Display the two clock images. Say, *These show time.* Emphasize that *time* ends with /m/. Then display the multiplication equation and the array. For each, say, *This shows ____ times ____.* Emphasize that *times* ends with /z/. Finally, display each of the four images in random order, and have students correctly identify it by saying **time** or **times**. | Display these sentence frames: 1. ____ **times as much is** ____. 2. ____ **times more is** ____. 3. ____ **times as many is** ____. Then ask, *What is 8 times as much as 4?* Have a volunteer write the corresponding equation, $4 \times 8 = 32$, and then answer your question using Sentence Frame 1: **8 times as much is 32**. Use this routine to have students answer the following questions, using Sentence Frames 2 and 3, respectively: *What is 5 times more than 6?* $5 \times 6 = 30$; **5 times more is 30**. *What is 2 times as many as 7?* $2 \times 7 = 14$; **2 times as many is 14.** | Display this sentence frame: ____ **times more than** ____ **is** ____. Divide students into pairs, and give each pair a number cube. Then have students use this routine to help them create multiplication sentences: First, one student rolls a number (for example, 2). Next, the other student rolls a number (for example, 5). Then both students use the two numbers rolled to create a multiplication sentence, using the sentence frame. (For example, **5 times more than 2 is 10.**) Have students repeat the routine several times. |

### Teacher Notes:

NAME _____ DATE _____

# Lesson 3 Concept Web
## *Multiplication as Comparison*

Use the concept web to write examples of ways to indicate 2 × a number.

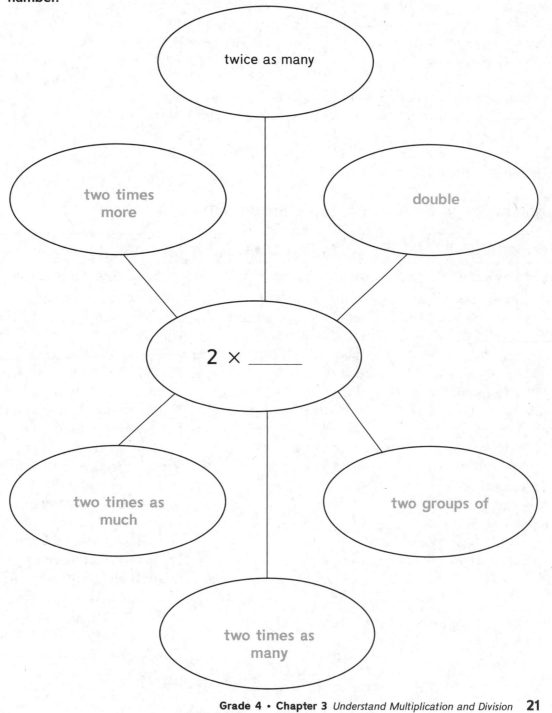

# Lesson 4 Compare to Solve Problems
## English Learner Instructional Strategy

### Vocabulary Support: Frontload Academic Vocabulary

Write *divide, multiply,* and *variable,* along with their Spanish cognates, *divider, multiplicar,* and *variable,* respectively, on a cognate chart. Introduce the words, and provide visual math examples. Utilize other appropriate translation tools for non-Spanish speaking ELs. Explain that *variable* is a multiple-meaning word that can mean changeable. *However, in math, variable means an unknown quantity.* Discuss how letters are used in math problems as variables.

Write *compare/comparison* on the board. Say, *When we **compare**, or make a **comparison**, we tell how things are the same.* Write this equation: $4 + 5 = 9$. Say, *One way to say this is, "Four plus five is **equal** to nine." Another way to say it is, "Four plus five is the **same** as nine."* Discuss *comparison* and *equation* as synonyms.

### English Language Development Leveled Activities

| Emerging Level | Expanding Level | Bridging Level |
|---|---|---|
| **Word Knowledge** | **Number Sense** | **Public Speaking Norms** |
| Write this equation, and then say: $d \times 4 = 24$. Point to the d and say, *This is a variable. The variable is d.* Have students repeat chorally, **The variable is d.** Then display this sentence frame: **The variable is ____.** Write other equations with letter variables, such as $3 \times e = 21$. Ask students to use the sentence frame to help them identify the variable in each. (For example: **The variable is e.**) | Write the following equation on the board: $5 + n = 27$. Have a volunteer come to the board to write and say the value of the variable. $n = 22$; **n equals twenty-two.** Repeat the exercise, using equations with different operations and letter variables. | Display the following: *There are 15 white jerseys, 12 blue jerseys, 3 yellow jerseys, and 4 green jerseys.* Have each student write a comparison question based on this information. (For example: **How many times as many blue jerseys are there than green jerseys?**) Then have partners exchange questions. Ask students to write an equation with a variable, based on their partners' comparison question, and then find the unknown. (For example: $4 \times b = 12$; $b = 3$; **There are 3 times as many blue jerseys.**) Have students present their responses. |

**Teacher Notes:**

NAME _____  DATE _____

# Lesson 4 Note Taking

## *Compare to Solve Problems*

Read the question. Write words you need help with and research each word. Use your lesson to write your Cornell notes. Write or draw math examples to explain your thinking. Share your examples with a classmate.

| **Building on the Essential Question** | **Notes:** |
|---|---|
| How can you compare to solve problems? | When you <u>compare</u> two values, you are identifying how they are the <u>same</u>. |
| | When a problem asks *how much more,* it is an <u>additive</u> comparison and requires <u>addition</u> or <u>subtraction</u> as the operation to compare. |
| | Other phrases that indicate additive comparison are how <u>many</u> more and how <u>much</u> less. |
| | When a problem asks *how many times more,* it is a <u>multiplicative</u> comparison and requires <u>multiplication</u> or <u>division</u> as the operation to compare. |
| **Words I need help with:** See students' words. | Another phrase that indicates multiplicative comparison is *how* <u>many</u> *times* <u>greater</u>. |
| | When solving comparisons, you can use an <u>equation</u> with a <u>variable</u> in place of the unknown value. |

**My Math Examples:**

See students' examples.

# Lesson 5 Multiplication Properties and Division Rules

## English Learner Instructional Strategy

### Vocabulary Support: Activate Prior Knowledge

Remind students that *property* is a multiple-meaning word, and ask what it means in a math context. **Property means "rule."** Then ask, *What are some Properties of Addition we have learned about?* **The Identity Property, Commutative Property, and Associative Property.** Use the My Vocabulary Cards from Chapter 2 to review. Then say, *In the coming lessons, we will learn about Properties of Multiplication. Like addition, multiplication has an Identity Property, a Commutative Property, and an Associative Property.* Explain how the memory devices students used with these properties in addition can also help them in multiplication. Make an anchor chart with the following: **The Identity Property is about a factor and product whose *identity* is the same. The Commutative Property is about factors that *commute*, or move/travel. The Associative Property is about factors that *associate*, or work together.** Have students read aloud with you.

### English Language Development Leveled Activities

| Emerging Level | Expanding Level | Bridging Level |
|---|---|---|
| **Word Knowledge** Write then say: $5 \times 1 = 5$. Point to each 5 and say, *These are the same. This shows the Identity Property.* Have students repeat, **Identity Property.** Next write $7 \times 0 = 0$, point to each 0, and say, *These are both zero. This shows the Zero Property.* Have students repeat, **Zero Property.** Finally, write $6 \times 2 = 12$ and $2 \times 6 = 12$. Point to the 2 and 6 in each equation, and say, *These moved places…* Then point to the 12 in both equations and say, *… but these are the same. This shows the Commutative Property.* Have students repeat, **Commutative Property.** | **Listen and Identify** Write this on the board: $14 \times 1 = 14$. Say, *This equation shows the Identity Property of Multiplication.* Explain how this property states that when any number is multiplied by 1, the product is that number. Then write this on the board: $0 \times 32 = 0$. Say, *This equation shows the Zero Property of Multiplication.* Explain how this property states that when any number is multiplied by 0, the product is 0. Write other equations for students to identify by saying **Identity Property of Multiplication** or **Zero Property of Multiplication.** | **Public Speaking Norms** Divide students into three multilingual groups. Assign each group one of the Properties of Multiplication from the lesson. Have each group prepare a presentation that explains their assigned property and gives examples. Have groups make their presentations to the class. Tell students to make sure each group member participates in the presentation. |

**Teacher Notes:**

NAME _____ DATE _____

# Lesson 5 Vocabulary Cognates
## *Multiplication Properties and Division Rules*

Use the Glossary to define the math word in English and in Spanish in
the word boxes. Write a sentence using your math word.

| Commutative Property of Multiplication | propiedad conmutativa de la multiplicación |
|---|---|
| **Definition** The property that states that the order in which two numbers are multiplied does not change the product. | **Definición** Propiedad que establece que el orden en el que se *multiplican* dos o más números no altera el *producto*. |
| **My math word sentence:** Sample answer: The Commutative Property of Multiplication states that 2 × 7 has a product that is equal to the product of 7 × 2. | |

| Identity Property of Multiplication | propiedad de identidad de la multiplicación |
|---|---|
| **Definition** If you *multiply* a number by 1, the *product* is the same as the given number. | **Definición** Si *multiplicas* un número por 1, el *producto* es igual al número dado. |
| **My math word sentence:** Sample answer: The Identify Property of Multiplication states that 8 × 1 equals 8. | |

| Zero Property of Multiplication | propiedad del cero de la multiplicación |
|---|---|
| **Definition** The property that states any number multiplied by zero is zero. | **Definición** Propiedad que establece que cualquier número *multiplicado* por cero es igual a cero. |
| **My math word sentence:** Sample answer: The Zero Property of Multiplication states that 0 × 5 equals 0. | |

# Lesson 6 The Associative Property of Multiplication

## English Learner Instructional Strategy

### Vocabulary Support: Activate Prior Knowledge

Write *product* and its Spanish cognate, *producto*. Introduce the words, and provide visual examples to support understanding. Utilize other appropriate translation tools for non-Spanish speaking ELs. Then tell students *product* is a multiple-meaning word. For example, it can mean "consequence" or "something offered for sale." Ensure students know that, in math, it means "the answer to a multiplication problem."

Then write these two sets of equations on the board:

$(4 + 1) + 3 = 8$      $(3 \times 2) \times 1 = 6$

$4 + (1 + 3) = 8$      $3 \times (2 \times 1) = 6$

Say, *Both sets of equations show the Associative Property.* Point to the plus signs in the first set and say, *This set shows the Associative Property of Addition.* Then point to the multiplication symbols in the second set and say, *This set shows the Associative Property of Multiplication.* Then ask, *How are all of the equations alike?* **All have parentheses.** Remind students that, in math, parentheses show which part of a problem to solve first.

### English Language Development Leveled Activities

| Emerging Level | Expanding Level | Bridging Level |
|---|---|---|
| **Academic Word Knowledge** | **Act It Out** | **Number Sense** |
| Write $4 \times 2 \times 3$ on the board. Demonstrate the following groupings: $(4 \times 2) \times 3$ and $4 \times (2 \times 3)$ Point out that, no matter how the numbers are grouped, the product is the same. Say, *This is a math rule. It is the Associative Property of Multiplication.* Have students recite the name of the property chorally at least three times. Then have students choose 3 single-digit numbers. Ask them to write two multiplication sentences, grouping the numbers differently for each. For example: **$(5 \times 1) \times 2$ and $5 \times (1 \times 2)$**. | Divide students into pairs. Distribute a number cube to each pair. Have partners roll the number cube three times and write them as an equation, for example: $5 \times 3 \times 4$. Then have one partner group the first two factors and then multiply them by the third factor, for example: $(5 \times 3) \times 4$. Next, have the other partner group the second and third factors and then multiply them by the first factor, for example: $5 \times (3 \times 4)$. Finally, have students compare solutions. Ask partners to present their work to the class using the terms: **parenthesis, factor, product**. | Have each student write a multiplication problem that contains three factors, such as **$8 \times 5 \times 10$**. Then ask students to suggest a way to group the factors in their problem, such as **$(8 \times 5) \times 10$**. Tell them that they should try the grouping that makes the problem easiest to solve. Finally, ask students to exchange papers with a partner and solve each other's problem. Have students check each other's work and then discuss whether the suggested order of factors affected how easy the problem was to solve. |

**Teacher Notes:**

NAME _____ DATE _____

# Lesson 6 Vocabulary Definition Map
## *The Associative Property of Multiplication*

Use the definition map to write a description and list characteristics about the vocabulary word or phrase. Write or draw math examples. Share your examples with a classmate.

My Math Vocabulary:

**Associative Property of Multiplication**

**Characteristics from Lesson:**

The Associative Property of <u>Addition</u> is similar to the Associative Property of Multiplication because they both have <u>parentheses</u>.

The <u>parentheses</u> let you know which two numbers to multiply <u>first</u>.

This property allows you to determine which <u>two</u> factors should be <u>multiplied</u> together first to make it easier to find the final <u>product</u>.

**Description from Glossary:**

The property that states that the grouping of the *factors* does not change the *product*.

**My Math Examples:**

Sample answers:

$3 \times (6 \times 2) = (3 \times 6) \times 2$

$8 \times (4 \times 7) = (8 \times 4) \times 7$

# Lesson 7 Factors and Multiples

## English Learner Instructional Strategy

### Vocabulary Support: Mnemonic Device

Before the lesson, write *factor* and *multiple*, along with their Spanish cognates, *factor* and *múltiplo*, on a cognate chart. Introduce the words, and provide visual examples to support understanding. Utilize other appropriate translation tools for non-Spanish speaking ELs.

Offer students this mnemonic device for the meaning of multiple: *Multiply to find a number's multiple.* As you say it, emphasize the /ī/ in *multiply* and the /əl/ in *multiple.* Then say, *The mnemonic device is true because a multiple is always the product of two numbers. What is a product?* **The answer to a multiplication problem.**

Then write *decompose* on the board. Tell students that this word has different meanings in different contexts. Say, *In science, decompose means "to decay," but in math it means "to break into smaller parts."* Provide illustrations or pictures to support understanding.

### English Language Development Leveled Activities

| Emerging Level | Expanding Level | Bridging Level |
|---|---|---|
| **Phonemic Awareness** | **Academic Vocabulary** | **Do the Math** |
| On the board, write $2 \times 7 = 14$. Point to the multiplication symbol and say, *multiply,* emphasizing the /ī/ at the end of *multiply.* Have students repeat **multiply** chorally. Then point to 14 in the highlighted row or column of a multiplication chart. Say, *14 is a multiple of 7,* emphasizing the /əl/ at the end of *multiple.* Have students repeat **multiple** chorally. Then point to 0 and 7 in the highlighted row or column of the multiplication chart. Say, *0 and 7 are also multiples of 7,* emphasizing the /z/ at the end of *multiples.* Have students repeat **multiples** chorally. | Draw the following chart on the board: <br><br> Guide students in recognizing that 1, 2, 3, and 6 are factors of 6, and 0, 6, 12, 18, and 24 are *multiples* of 6. Then make a chart for 8, listing its factors (1, 2, 4, 8) and first five multiples (0, 8, 16, 24, 32). Have students use these sentence frames to distinguish the sets of numbers: _____ **are factors of** _____. _____ **are multiples of** _____. | Have each student draw an array and then exchange papers with a partner. Tell students to identify the total number of objects in the array and then find the factors for that number. (For example, if the array has 8 objects, then students should find the factors for 8, which are 1, 2, 4, and 8.) Finally, have students find the first five multiples of the number. (For example: The first five multiples of 8 are 0, 8, 16, 24, and 32.) Have partners check each other's work. |

Chart within Expanding Level:

| 6 | |
|---|---|
| 1,2,3,6 | 0,6,12,18,24 |

**Teacher Notes:**

NAME _____ DATE _____

# Lesson 7 Four-Square Vocabulary
## *Factors and Multiples*

Write the definition for each math word. Write what each word means in your own words. Draw or write examples that show each math word meaning. Then write your own sentences using the words.

| **Definition** | **My Own Words** |
|---|---|
| A multiple of a number is the *product* of that number and any whole number. | See students' examples. |

**multiple**

| **My Examples** | **My Sentence** |
|---|---|
| Sample answer: 15 is a multiple of 5 because 3 × 5 = 15 | Sample sentence: The first five multiples of 4 are 0, 4, 8, 12, 16. |

| **Definition** | **My Own Words** |
|---|---|
| To break a number into different parts. | See students' examples. |

**decompose**

| **My Examples** | **My Sentence** |
|---|---|
| Sample answer: The number 20 can be decomposed into 4 × 5. | Sample sentence: I can break down or decompose the number 15 into its factors, 5 × 3. |

# Lesson 8 Problem-Solving Investigation
## Strategy: Reasonable Answers
### English Learner Instructional Strategy

**Language Structure Support: Tiered Questions**

Before the lesson, remind students of the meaning of *reasonable*. Then ask, *If it is summertime and the sun is shining, is it reasonable to guess that it is cold outside?* **No** *Why not?* **It is usually warm or hot during the summer.**

During Practice the Strategy, ask these tiered questions to assist ELs with Steps 1–3: *Some facts we know are the numbers of newspapers that Dasan and Lisa deliver. How many newspapers does Dasan deliver?* **283** *How many newspapers does Lisa deliver?* **302** *Now, what do we need to find?* **The number of newspapers Dasan and Lisa deliver together.** *Let's make a plan for finding this number. Should we add or subtract?* **Add** *Do we need to find an exact number?* **No** *How do you know?* **The problem asks if 400 is a reasonable estimate.** *So, let's round. What is 283 rounded to the nearest hundred?* **300** *What is 302 rounded to the nearest hundred?* **300** *What is two times as many as 300?* **600** *So, was "400 newspapers" a reasonable estimate?* **No**

## English Language Development Leveled Activities

| Emerging Level | Expanding Level | Bridging Level |
|---|---|---|
| **Word Knowledge** | **Internalize Language** | **Show What You Know** |
| Say, *When something is reasonable, it makes sense.* Then write the following: 9 + 5 > 12; 20 × 4 < 50. Work through the first inequality with students, adding 9 + 5 to find 14. Say, *14 is greater than 12. This is reasonable.* Prompt students to say, **Reasonable.** Then work through the second inequality with students, multiplying 20 × 4 to find 80. Say, *80 is not less than 50. This is not reasonable.* Prompt students to say, **Not reasonable.** Have students identify other inequalities as **Reasonable** or **Not Reasonable.** | Draw a group of 8 hearts labeled "8", a group of 9 hearts labeled "9", and a group of 11 hearts labeled "11". Ask, *Is it reasonable to say there are about 30 hearts in all?* Guide students in rounding 8, 9, and 11 to 10 and then adding 10 + 10 + 10 to get an estimate of 30. Then add 8 + 9 + 11 to get 28. Say, *The estimate is reasonable because 30 is close to 28.* Discuss what an unreasonable estimate would be. Then display another set of grouped objects and estimate its number. Have students explain why the estimate is reasonable or unreasonable. | Assign each student a number. Have students write a word problem for which their assigned number is a reasonable answer. Then ask students to present their numbers and problems. After each presentation, have the rest of the class solve the problem to check the presenter's work. |

**Teacher Notes:**

NAME _____ DATE _____

# Lesson 8 Problem-Solving Investigation
## STRATEGY: Reasonable Answers

Determine a reasonable answer to each problem.

| Pennies Collected | |
|---|---|
| Child | Number of Pennies |
| Myron | 48 |
| Teresa | 52 |
| Veronica | 47 |
| Warren | 53 |

1. The **table** shows the number of **pennies collected** by **four children.**

   Is it reasonable to say that **Myron** <u>and</u> **Teresa** collected <u>**about**</u> 100 pennies in <u>**all**</u>? Explain.

| Understand | Solve |
|---|---|
| I know: <br><br><br> I need to find: | |
| **Plan** <br><br> 48 rounds to... <br><br> 52 rounds to... | **Check** |

2. **Jay** will make <u>$240</u> doing yard work for **6 weeks.**
   **He** (Jay) is **saving** his money to **buy** camping equipment that <u>costs $400.</u>
   **He has** already saved <u>$120.</u>
   Is it **reasonable** to say that Jay will <u>**save enough**</u> money **in 6 weeks**? Explain.

| Understand | Solve |
|---|---|
| I know: <br><br><br> I need to find: | |
| **Plan** <br><br> ⌐--------------- ? ---------------⌐ <br> \| 120 \| 240 \| | **Check** |

# Chapter 4 Multiply with One-Digit Numbers

## What's the Math in This Chapter?

### Mathematical Practice 7: Look for and make use of structure.

Write $4 \times 6,000,000 = ?$ on the board. Display a demonstration place-value chart and ask, *How can we use place value to solve this problem? Turn and talk with a friend to discuss my question.*

Have student share their ideas. Introduce the concept of using the basic fact $4 \times 6$ if it is not mentioned. Ask, *What is the basic fact $4 \times 6$?* **24** Write 24 in the place-value chart. Ask, *What is $4 \times 60$?* **240** Write 240 below 24 in the place-value chart. Repeat with each incremental factor until you find the product of $4 \times 6,000,000$. **24,000,000**

Ask, *How did the **structure** of place value help us solve the problem?* Allow students time to share their ideas. The goal is for them to see that all multiplication is based on place value. It is important for students to make the connection so they will transition more easily into multiplication with greater numbers.

Display a chart with Mathematical Practice 7. Restate Mathematical Practice 7 and have students assist in rewriting it as an "I can" statement, for example: **I can understand multiplication by thinking about place value.** Have students write examples of using place value to multiply. Post the examples and the new "I can" statement.

## Inquiry of the Essential Question:

### How can I communicate multiplication?

Inquiry Activity Target: **Students come to a conclusion that multiplication uses place value.**

As an introduction to the chapter, present the Essential Question to students. The inquiry graphic organizer will offer opportunities for students to observe, make inferences, and apply prior knowledge of place value representing the Essential Question. As they investigate, encourage students to draw, write, and collaborate with peers to demonstrate their observations and thinking. Then have students present additional questions they may have to a peer to extend discussions.

Regroup students and restate Mathematical Practice 7 and the Essential Question. Pose questions to reflect on what has been learned to guide students in making connections between the Mathematical Practice and the Essential Question.

NAME _____ DATE _____

# Chapter 4 Multiply with One-Digit Numbers

## Inquiry of the Essential Question:

**How can I communicate multiplication?**

**Read the Essential Question. Describe your observations (I see...), inferences (I think...), and prior knowledge (I know...) of each math example. Write additional questions you have below. Then share your ideas and questions with a classmate.**

$6 \times 7 = 42$      Basic fact             I see ...

$6 \times 70 = 420$      $6 \times 7$ tens $= 42$ tens

$6 \times 700 = 4{,}200$    $6 \times 7$ hundreds $= 42$ hundreds     I think...

I know...

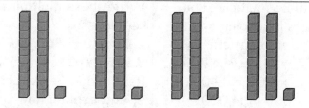

I see ...

I think...

I know...

Number of tens: 8

Number of ones: 4

8 tens + 4 ones = 80 + 4 = 84

$4 \times 21 =$ _____

$26 \times 7 = (20 \times 7) + (6 \times 7)$ Write 26 as 20 + 6.    I see ...

$= \quad 140 \quad + \quad 42$    Multiply mentally.
Add.

$= \qquad 182$           I think...

I know...

Questions I have...

_____

_____

_____

# Lesson 1 Multiples of 10, 100, and 1,000
## English Learner Instructional Strategy

### Sensory Support: Manipulatives/Photos

Before the lesson, review the meaning of *multiple*. Write 6: 0, 6, 12, 18, 24, 30, 36, 42, 48, 54 . . . Ask, *What are the numbers listed after the 6 called?* **multiples** Ask, *How are the words **multiple** and **multiply** related?* Discuss all student responses. Write, then have students say chorally, **You must multiply a number to find its multiples.**

As you guide students through the Math in My World examples, mode using base-ten blocks to help them visualize equations that include the words *tens*, *hundreds*, and *thousands*.

Use photos/realia to review English terms from the Problem Solving exercises, which may be unfamiliar, such as: *ticket, theme park, cost, week,* and *rides*.

### English Language Development Leveled Activities

| Emerging Level | Expanding Level | Bridging Level |
|---|---|---|
| **Show What You Know** | **Show What You Know** | **Developing Oral Language** |
| Write these multiplication sentences in a column to show the zero pattern: 1. $3 \times 6 = 18$; 2. $3 \times 60 = 180$; 3. $3 \times 600 = 1,800$; 4. $3 \times 6,000 = 18,000$. Point to the 6 in Sentence 1, and say, *There is no zero. $3 \times 6$ is basic fact.* Then point to 18 in the same sentence, and say, *There is no zero.* Next, point to 60 in Sentence 2 and say, *There is one zero.* Then point to 180 in the same sentence, and say, *There is one zero.* Guide students in continuing this process with Sentences 3 and 4. Have them use this sentence frame for help: **There are _____ zeros.** | Write these multiplication sentences in a column to show the zero pattern: 1. $2 \times 7 = 14$; 2. $2 \times 70 = 140$; 3. $2 \times 700 = 1,400$; 4. $2 \times 7,000 = 14,000$. Guide students in comparing the sentences and recognizing that in each, the second factor and the product have the same number of zeros. Then write these multiplication sentences: 5. $6 \times 9 = $ _____; 6. $6 \times 90 = $ _____; 7. $6 \times 900 = $ _____; 8. $6 \times 9,000 = $ _____. Divide students into small groups. Then have them find the unknowns in Sentences 5–8, using the number pattern in Sentences 1–4 for help. | Write these multiplication sentences: 1. $7 \times 8 = 56$; 2. $7 \times 80 = 560$; 3. $7 \times m = 5,600$; 4. $7 \times m = 56,000$; Read aloud Sentences 1 and 2, and explain how the two sentences are related. Then read aloud Sentences 3 and 4. Have students Turn and Talk with a peer about the number pattern started in Sentences 1 and 2. Finally, have students use the number pattern to help them find the unknowns in Sentences 3 and 4. Ask them to share their answers with the group, using this sentence frame: **If $7 \times 8 = 56$, then $7 \times$ _____ equals _____.** |

### Multicultural Teacher Tip

Mathematical notation varies from culture to culture, so you may find ELs using unfamiliar symbols in place of standard US symbols. For example, students from Latin American countries may use a point in place of $\times$ to indicate multiplication. Although the point is also commonly used in the US, the placement and size may vary depending on the student's native culture. In Mexico, the point is larger and set higher between the numbers than in the US. In some Latin American countries, the point is set low and can be confused with a decimal point.

NAME _____ DATE _____

# Lesson 1 Concept Web
## *Multiples of 10, 100, and 1,000*

Use the concept web to write examples of multiplying using multiples and basic facts.

$4 \times \underline{70} = 280$

$4 \times \underline{700} = 2,800$

$4 \times \underline{7,000} = 28,000$

$4 \times 7 = 28$
$7 \times 4 = 28$

$7 \times \underline{40} = 280$

$7 \times \underline{400} = 2,800$

$7 \times \underline{4,000} = 28,000$

# Lesson 2 Round to Estimate Products
## English Learner Instructional Strategy

### Vocabulary Support: Academic Vocabulary

Before the lesson, write *digit* and *multiply* and their Spanish cognates, *dígito* and *multiplicar* on a cognate chart. Introduce the words, and provide a visual math example to support understanding. Utilize other appropriate translation tools for non-Spanish speaking ELs. Discuss both math and non-math meanings for the words.

Before teaching the Math in My World examples, present the following chant to assist with rounding: **4 down to 0, round down. 5 up to nine, round up.** Have students say chorally several times.

When discussing the value of the actual product, use the following gestures: Point up when saying, *Rounding up makes the estimate higher than the product.* Point down when saying, *Rounding down makes the estimate lower than the product.*

### English Language Development Leveled Activities

| Emerging Level | Expanding Level | Bridging Level |
|---|---|---|
| **Academic Vocabulary** | **Graphic Support** | **Number Game** |
| Write the number 486 on the board. Point to each digit from right to left and say, *Ones, tens, hundreds,* as you point to the appropriate place value. Now underline the 4 and say, *The greatest place value is hundreds,* emphasizing the word *greatest.* Have students repeat the sentence chorally. On the board, write random numbers consisting of two to four digits. Have students use the following sentence frame to identify the greatest place value: **The greatest place value is _____.** | Write $146 \times 2 = $ _____ and $756 \times 4 = $ _____ on the board. Pair students. Provide each pair with a number line in increments of 100 and another in increments of 1000. Have pairs estimate both products, use a calculator to find the exact products, and plot both on the appropriate number line. Have students determine if the estimate is greater or less than the product. Students report back using the number lines to show: **When we round _____, the estimate is greater than the actual product, and when we round _____, the estimate is less than the actual product.** | Divide students into small multilingual groups, and distribute a number cube to each group. Have students in each group take turns rolling the number cube to establish the numbers for a one-digit by three-digit multiplication problem. Once the problem has been set up for all group members to see, students race to round to the greatest place and estimate the product. The first student to correctly determine the estimate wins that round. Have group members check each other's work. |

### Teacher Notes:

NAME _____ DATE _____

# Lesson 2 Multiple Meaning Word
## *Round to Estimate Products*

Complete the four-square chart to review the multiple meaning word or phrase.

| | |
|---|---|
| **Everyday Use**<br><br>Sample answer: Shaped like a circle or sphere. A basketball is round. | **Math Use in a Sentence**<br><br>Sample sentence: 327 rounded to the nearest 10 is 330. |
| **Math Use**<br><br><br>To change the value of a number to one that is easier to work with. To find the nearest value of a number based on a given *place value*. | **Example From This Lesson**<br><br><br>Sample answer: 327 rounded to the greatest place value is 300. |

**round**

Complete the sentences below by writing the correct numbers or terms on the lines.

To find the estimate of 917 × 3, round 917 to the greatest place value which is the ____hundreds____ place.

Then multiply ____900____ × 3 to find the estimate ____2,700____.

# Lesson 3 Inquiry/Hands On: Use Place Value to Multiply

## English Learner Instructional Strategy

### Sensory Support: Realia and Pictures

Before the lesson, write and say, *set*. Have students chorally repeat. Explain that *set* is a multiple-meaning word with a number of noun, verb, and adjective meanings. Invite students to share any meanings they know. (For example, *set* can be a noun meaning "exercises done in a series," a verb meaning "to place in a certain position," or an adjective meaning "unmoving.") Then tell students that in this lesson, you will use *set* to mean "a group of objects that belong together." Reinforce this meaning by showing realia and pictures of items that come in sets, such as books, tools, games, and so on.

Next, tell students that in this lesson they will be using base-ten models to make sets. Begin the Build It activity. Write 3 × 23 on the board and say, *This expression shows that we need 3 **sets** of 23 base-ten models.* Ask students to tell you how to make one set of 23. Then show how 3 sets of 23 are modeled in the Student Edition on page 209.

### English Language Development Leveled Activities

| Emerging Level | Expanding Level | Bridging Level |
| --- | --- | --- |
| **Listen and Identify** | **Cooperative Learning** | **Show What You Know** |
| Display a few ones models and a couple tens models. Point to the ones models and say, *These are **ones**. There is only **one** block in a **ones** model. **Ones**.* Have students chorally repeat. Then point to the tens models and say, *These are **tens**. There are **ten** blocks in a **tens** model. **Tens**.* Have students chorally repeat. Next, write: 22. Ask, *How many **ones** are in 22?* **two** *How many **tens** are in 22?* **two** Show how to represent 2 tens and 2 ones with the base-ten models. Repeat the routine using another 2-digit number, this time having students tell you how to represent each number using the base-ten models. | Pair up students for Practice It Exercises 7 and 8. Assign each pair one problem to work on together. Have students use this communication guide to help them discuss their problem: **The problem ____ × ____ shows that we need ____ sets of ____ blocks. One set has ____ tens and ____ ones. ____ sets have a total of ____ ones and ____ tens. ____ ones plus ____ tens equals ____. So, [variable] equals ____.** Then have partners meet with a pair of students who worked on a different problem. Have each pair explain their problem to the other pair. | **Show What You Know** Ask students to recall what it means to *decompose* a number. **break it into parts** Then pair up students and assign Apply It Exercise 12. Have partners work together to solve the problem, using this communication guide to help them as they discuss: **_a_ ____ _b_ = 88. _a_ ____ _b_ = 26. We can decompose 88 into ____ sets of ____. If _a_ equals ____ and _b_ = ____, then the product of _a_ and _b_ is ____ and the sum of _a_ and _b_ is ____.** Once the solution is found, ask them to teach the process to an Emerging or Expanding Level peer. |

## Teacher Notes:

NAME _____ DATE _____

# Lesson 3 Guided Writing

## *Inquiry/Hands On: Use Place Value to Multiply*

**How do you use place value to multiply?**

**Use the exercises below to help you build on answering the Essential Question. Write the correct word or phrase on the lines provided.**

**1.** Rewrite the question in your own words.

See students' work.

_____

**2.** What key words do you see in the question?

place value, multiply

**3.** Identify the number that is modeled below.

41

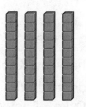

**4.** To model the multiplication equation 2 × 41, you would model the number __41__ using base ten blocks __two__ times.

**5.** What multiplication **equation** is modeled below?

2 × 24

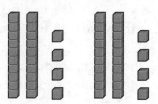

**6.** Use the modeled multiplication equation above to identify the total number of tens and ones. Record in the place-value chart.

| tens | ones |
|:----:|:----:|
| 4 | 8 |

**7.** How do you use place value to multiply?

I can use base-ten blocks to show the numbers in each group. Then I
can write the number of ones and tens in a place-value chart.

**30** **Grade 4 · Chapter 4** *Multiply with One-Digit Numbers*

# Lesson 4 Inquiry/Hands On: Use Models to Multiply

## English Learner Instructional Strategy

### Vocabulary Support: Build Background Knowledge

Before the lesson, write and say, *partial product*. Then underline the letters *part* in *part*ial and say, *A partial product is **part** of a product*. Write: *part, whole*. Discuss how these words are opposites. Reinforce meaning by showing students pictures of whole objects that are made up of parts or divided into parts, such as a pocket watch whose inner workings are exposed or a pie cut into wedges. As you show each object, gesture around it and say, *This is the **whole** ____.* Point to each of its parts and say, *This is **part** of the ____.* Gesture toward all the parts and say, ***Together**, all these **parts** make a **whole** ____.* Then randomly point out a whole object or part of the object and ask, *Is this the **whole** ____ or **part** of the ____?* Have students respond by saying **part** or **whole**.

Finally, return to the term partial product. Ask, *If a **partial** product is **part** of a product, should you **add** or **subtract** to make a **whole** product?* **add**

## English Language Development Leveled Activities

| Emerging Level | Expanding Level | Bridging Level |
|---|---|---|
| **Phonemic Awareness** | **Communication Guide** | **Show What You Know** |
| Write *partial product*. In *partial*, draw a vertical line between the *r* and *t* to divide the word into syllables. Point to the first syllable and say, */par/.* Have students chorally repeat. Point to the second syllable and say, */shəl/.* Have students chorally repeat. Show that the letters *ti* say /sh/ rather than /ti/ or /tī/. Write and say other words with *ti* pronounced /sh/, such as *multiplication* and *addition*. Then give students practice saying words with /sh/, such as **show, should, shower**, and **shot**. Demonstrate how /sh/ is different from /s/ or /ch/. | Remind students that a partial product is *part* of a whole product. Then pair up students to work on Practice It Exercises 3–7. Have students use the following communication guide to help them visualize the arrays and discuss the solutions: ____ **times** ____ **tens equals part of the product:** ____. ____ **times** ____ **ones equals part of the product:** ____. ____ + ____ **= the whole product:** ____. After students have completed their work, have them check answers with another set of partners. | Have students do the Expanding Level activity, using these sentence frames instead: **Multiply ____ by ____ tens to find one part of the product: ____ × ____ = ____. Multiply ____ by ____ ones to find the other part of the product: ____ × ____ = ____. Add the partial products to find the whole product: ____ + ____ = ____.** When they have completed their work, ask students to draw one of the arrays from their Practice It Exercises on a new sheet of paper, and then draw beside it a base-ten model that shows the same equation. Post drawings in the classroom. |

## Teacher Notes:

NAME _____ DATE _____

# Lesson 4 Four-Square Vocabulary

## Inquiry/Hands On: Use Models to Multiply

Write the definition for each math word. Write what each word means in your own words. Draw or write examples that show each math word meaning. Then write your own sentences using the words.

| | |
|---|---|
| **Definition** | **My Own Words** |
| A multiplication method in which the products of each place value are found separately, and then added together. | See students' examples. |

### partial products

| | |
|---|---|
| **My Examples** | **My Sentence** |
| Sample answer: The partial products of 7 × 39 are 210 (7 × 30) and 63 (7 × 9). | Sample sentence: You can add partial products to find the product of a two-digit number and a one-digit number. |

| | |
|---|---|
| **Definition** | **My Own Words** |
| The answer or result of a *multiplication* problem. It also refers to expressing a number as the *product* of its factors. | See students' examples. |

### product

| | |
|---|---|
| **My Examples** | **My Sentence** |
| Sample answer: The product of 7 and 3 is 21. | Sample sentence: The product of 4 × 4 is the same as the product of 8 × 2. Both products are equal to 16. |

Grade 4 • **Chapter 4** *Multiply with One-Digit Numbers* **31**

# Lesson 5 Multiply by a Two-Digit Number

## English Learner Instructional Strategy

### Vocabulary Support: Frontload Academic Vocabulary

Before the lesson, review the term *partial products*. Write *partial*, and then underline its base word, *part*. Say, *A partial product shows* **part** *of the product*. Write this problem on chart paper: 23 × 7 = _____ and draw an area model similar to the one in Example 1. Label the partial products and product. Briefly discuss. Keep posted in the classroom for students to reference during the lesson.

Use photos/realia to review English terms from the Problem Solving exercises, which may be unfamiliar, such as: *city, swing set, world's largest rodent, capybara, recycle, recycled, recycling,* and *action figures*.

### English Language Development Leveled Activities

| Emerging Level | Expanding Level | Bridging Level |
|---|---|---|
| **Academic Vocabulary** | **Listen and Identify** | **Public Speaking Norms** |
| Write the following equation on the board in vertical format: 12 × 4 = _____. Point to the 2 and say, *Multiply the ones digit.* Have students repeat chorally, and then write an 8 in the ones place of the product. Next, point to the 1 and say, *Multiply the tens digit.* Have students repeat chorally, and then write a 4 in the tens place of the product. Repeat the exercise with similar multiplication problems, having the students multiply by a two-digit number. | Write 2 × 24 = _____ in vertical format. Point to the 24 and say, *First, we multiply the ones. What digit is in the ones place?* **4 is in the ones place.** Multiply 4 by 2. Next point to the 24 again and say, *Now we multiply the tens. What digit is in the tens place?* **2 is in the tens place.** Finish multiplying to solve the problem. Repeat the exercise with students completing similar problems. Have students use the following sentence frame to identify ones and tens: **The digit in the ones/ tens place is _____.** | Divide students into multilingual groups. Assign each group one of the My Homework Practice exercises from the lesson. Have each group create an organized list explaining the steps of one-digit by two-digit multiplication. Direct students to use the assigned exercise as an example. Then ask groups to present their organized list to the class. Remind presenters to use good public speaking techniques, such as speaking clearly and making eye contact. |

**Teacher Notes:**

NAME _____ DATE _____

# Lesson 5 Note Taking

## *Multiply by a Two-Digit Number*

**Read the question. Write words you need help with and research each word. Use your lesson to write your Cornell notes. Write or draw math examples to explain your thinking. Share your examples with a classmate.**

| Building on the Essential Question | Notes: |
|---|---|
| **Building on the Essential Question**<br><br>How do you multiply by a two-digit number? | When you multiply a two-digit number by a one-digit number, first multiply the __ones__ place of the two-digit number. Then multiply the __tens__ place of the two-digit number.<br><br>For the number 34, the tens place value is __3__ and the ones place value is __4__.<br><br>For example:<br> 34<br> $\times 2$<br><br>First, multiply the ones place. 2 × __4__ ones = __8__ ones<br><br>Multiply the tens place. 2 × __3__ tens = __6__ tens |
| **Words I need help with:**<br><br>See students' words. | The sum of the partial products is the product.<br><br>__6__ tens and __8__ ones is equal to __68__.<br><br>34 × 2 = __68__ |

**My Math Examples:**

See students' examples.

# Lesson 6 Inquiry/Hands On: Model Regrouping

## English Learner Instructional Strategy

### Vocabulary Support: Word Knowledge

Before the lesson, write *regroup* and its Spanish cognate, *reagrupar*, on a cognate chart. Have students write the term *regroup* and its translation in their native language on an index card.

Next, cover up the letters *re* in *regroup* and say, *group*. Have students chorally repeat. Explain that objects are in a group when they are placed together. Ask volunteers to point out different examples of classroom objects that are placed in a group. For each example, say, *This is a **group** of _____.* Have students chorally repeat.

Finally, return to the word *regroup*, this time underlining the letters *re*. Explain that *re-* is a word part meaning "again" and that together *re-* and *group* spell a word meaning "group again." Use this as your lead-in for the Build It activities on Student Edition page 229. Have students draw examples of grouping and regrouping on their index cards for *regroup* and then store the cards in their math journals for future use.

### English Language Development Leveled Activities

| Emerging Level | Expanding Level | Bridging Level |
| --- | --- | --- |
| **Frontload Academic Vocabulary** | **Signal Words** | **Number Sense** |
| Write *exchange* and draw a vertical line between the *x* and *c* to divide the word into syllables. Point to the first syllable and say, /eks/. Have students chorally repeat. Point to the second syllable and say, /chānj/. Have students chorally repeat. Demonstrate how /sh/ is different from /ch/ and /j/ is different from /sh/ or /zh/. Finally, say the whole word together with students. Then have volunteers help you model using various items. Say, *I have _____. You have _____. Let's **exchange**. I will give you my _____. Now, you give me your _____.* | Have students do the Emerging Level activity. Display the following sentence frames and use them to discuss the synonym: *trade* and signal words: *in return.* **1) Let's exchange _____ for _____. 2) Let's trade _____ for _____. 3) I will give you _____. In return, you will give me _____.** After you have demonstrated how the sentence frames work, pair up students and ask them to write about an exchange of objects using each sentence frame to tell about the same exchange. Have students share their completed sentence frames with the class. | Pair up students and give each pair a number cube and a set of base-ten blocks. Have partners roll the number cube three times, recording each number rolled. Then ask them to use the numbers to make a three-digit number. Next, have them work together to model the number using only tens and ones blocks. Then have them model the number again, this time exchanging tens blocks for hundreds blocks, as appropriate. Finally, ask them to draw a poster that shows their number along with both models and have them present it to the class. |

**Teacher Notes:**

NAME _____ DATE _____

# Lesson 6 Definition Map
## *Inquiry/Hands On: Model Regrouping*

Use the definition map to write a description and list characteristics about the vocabulary word or phrase. Write or draw math examples. Share your examples with a classmate.

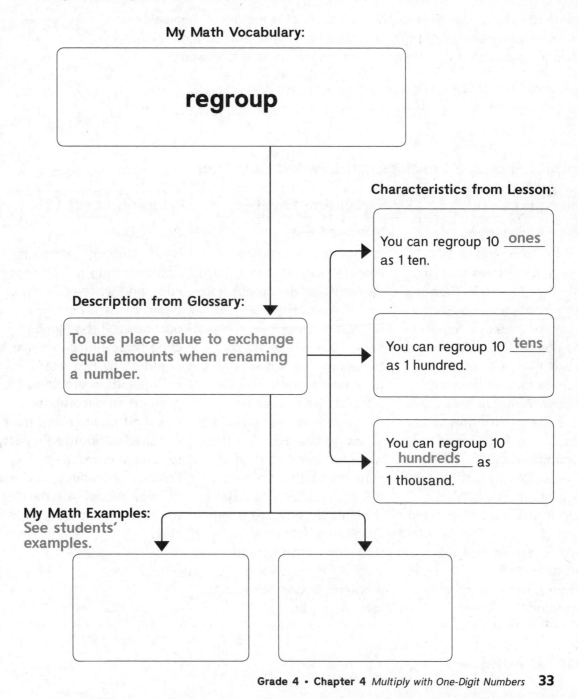

**My Math Vocabulary:**

**regroup**

**Characteristics from Lesson:**

You can regroup 10 <u>ones</u> as 1 ten.

**Description from Glossary:**

To use place value to exchange equal amounts when renaming a number.

You can regroup 10 <u>tens</u> as 1 hundred.

You can regroup 10 <u>hundreds</u> as 1 thousand.

**My Math Examples:**
See students' examples.

# Lesson 7 The Distributive Property

## English Learner Instructional Strategy

### Vocabulary Support: Build Background Knowledge

As an introduction to the lesson, stand in front of the class with a stack of paper. Say, *I have a stack of paper. Now I will* **distribute** *my stack of paper.* Model giving out paper to several students. As you approach each student, say, *I will distribute* **part** *of the paper to you.* When all of your paper is gone, say, *Now each of you has* **part** *of my stack of paper.* Explain that to distribute is to give out parts of something. Then write *Distributive Property of Multiplication* on the board, and underline *Distribut-* in *Distributive.* Briefly discuss *distribute* as a root word.

Say, *In this lesson we will learn about the Distributive Property of Multiplication. This rule involves breaking a number into* **parts** *before multiplying it.*

### English Language Development Leveled Activities

| Emerging Level | Expanding Level | Bridging Level |
|---|---|---|
| **Word Knowledge** | **Number Sense** | **Act It Out** |
| Hold up a strand of 14 connecting cubes and say, *Here are 14 cubes.* Then say, *Now I will* **decompose** *the group of 14 cubes.* Separate the connecting cubes into a set of 10 and a set of 4. Explain that *decompose* means "to break apart or separate." Write this on the board to help illustrate decomposition: 14 = 10 + 4. Then write random two-digit numbers (without a zero in the ones place) and have students demonstrate how to decompose them into tens and ones that can be used in an addition equation. | Review the Distributive Property with students. Then demonstrate decomposition. Say, *To decompose 46 into 40 plus 6 write it like this:* 46 = 40 + 6. Explain that to decompose a number is to separate it into smaller addends. Remind students that *addends* are numbers added to make a sum. Then write random two-digit numbers (without a zero in the ones place) and have students decompose them. Provide the following sentence frame to help students discuss each problem: **Decompose _____ into _____ plus _____.** | Divide students into pairs, and distribute a 0–5 number cube and write-on/wipe-off board to each pair. Have partners roll the number cube three times to make a one-digit by two-digit multiplication problem. Tell partners to decompose the two-digit number and then use the Distributive Property to solve the problem. Encourage students to draw an area model. Ask partners to present their work to the class. |

**Teacher Notes:**

NAME _____ DATE _____

# Lesson 7 Vocabulary Cognates
## *The Distributive Property*

Use the Glossary to define the math word in English and in Spanish in the word boxes. Write a sentence using your math word.

| Distributive Property | propiedad distributiva |
|---|---|
| **Definition** <br> To *multiply* a *sum* by a number, multiply each *addend* by the number and *add* the products. | **Definición** <br> Para *multiplicar* una *suma* por un número, multiplica cada *sumando* por ese número y luego *suma* los *productos*. |
| **My math word sentence:** <br><br> Sample answer: To find 4 × 15, think of 5 as 10 + 5. The distributive property states that 4 × (10 + 5) = (4 × 10) + (4 × 5). | |

| decompose | descomponer |
|---|---|
| **Definition** <br> To break a number into different parts. | **Definición** <br> Separar un número en diferentes partes. |
| **My math word sentence:** <br> Sample answer: The number 15 can be decomposed into 10 + 5. | |

# Lesson 8 Multiply with Regrouping
## English Learner Instructional Strategy

### Language Structure Support: Sentence Frames

Write *regroup* and its Spanish cognate, *reagrupar*. Introduce the word, and provide a visual math example to support understanding. Compare the math meaning of *regroup* with this slightly different nonmath meaning: "to organize anew after a difficult time." Say the following as an example of the word's use in a non-math context: *After losing several games, our team decided to regroup. Now that we have a new plan, we feel more confident.* To clarify meaning, have students utilize appropriate translation tools.

Post these sentence frames on an Anchor Chart for students to use as they respond to the Talk Math question: **First, multiply the ones: _____ × _____ = _____. Then regroup _____ ones as _____ tens and _____ ones. Next, multiply the tens: _____ × _____ = _____. Last, add the regrouped tens: _____ + _____ = _____.** Students can reference these sentence frames throughout the exercises.

### English Language Development Leveled Activities

| Emerging Level | Expanding Level | Bridging Level |
|---|---|---|
| **Exploring Language Structure**<br>Explain that *re-* is a word part meaning "again." Then write: *build/rebuild.* Demonstrate building, knocking down, and then rebuilding a block tower. As you demonstrate, say, *First, I build.* Then I *rebuild.* Next, write: *write/rewrite.* Demonstrate writing a word, erasing it, and then rewriting it. As you demonstrate, say, *First, I write. Then I rewrite.* Finally, write: *group/regroup.* Demonstrate dividing 16 connecting cubes among two groups of 8. Say, *First I group.* Then transfer 2 cubes from one train to the other. Say, *Then I regroup.* | **Act It Out**<br>Write $17 \times 3 =$ _____ in vertical format. Then distribute base-ten blocks to students and have them mimic as you model forming three groups of 17 and then regrouping to form five groups of 10 and one group of 1. As you demonstrate, pause at each step to ask, *Should I regroup?* Repeat the activity with similar multiplication problems that require regrouping. | **Show What You Know**<br>Divide students into multilingual groups. Ask each group to discuss when regrouping is necessary in multiplication problems. Then have each group prepare two multiplication problems, one that does not require regrouping and another that does. Have groups present their problems to the class, along with an explanation of why regrouping was or was not required in each. |

**Teacher Notes:**

NAME _____ DATE _____

# Lesson 8 Note Taking

## *Multiply with Regrouping*

**Read the question. Write words you need help with and research each word. Use your lesson to write your Cornell notes. Write or draw math examples to explain your thinking. Share your examples with a classmate.**

| Building on the Essential Question | Notes: |
|---|---|
| **How do you multiply with regrouping?** | When you multiply a two-digit number by a two-digit number, first multiply the <u>ones</u> place. Then <u>regroup</u> the ones into <u>tens</u> and <u>ones</u>. Finally multiply the <u>tens</u> place.<br><br>For the number 34, the tens place value is <u>3</u> and the ones place value is <u>4</u>.<br><br>For example:<br>34<br><u>×6</u><br><br>First, multiply the ones place. 6 × <u>4</u> ones = <u>24</u> ones.<br><br><u>24</u> ones are equal to <u>2</u> tens and <u>4</u> ones.<br><br>Next, multiply the tens place. 6 × <u>3</u> tens = <u>18</u> tens.<br><br>Add the tens that were regrouped after multiplying the ones.<br><u>18</u> tens + <u>2</u> tens = <u>20</u> tens.<br><br>The sum of the partial products is the product. <u>20</u> tens and <u>4</u> ones is equal to <u>204</u>.<br><br>34 × 6 = <u>204</u> |
| **Words I need help with:** See students' words. | |

**My Math Examples:**
See students' examples.

# Lesson 9 Multiply by a Multi-Digit Number

## English Learner Instructional Strategy

### Vocabulary Support: Build Background Knowledge

During Math in My World, Example 1, point out the term *leap years* in the word problem. Explain that a leap year occurs every four years and that, during a leap year, an extra day is added to the month of February. Say, *This makes a leap year 366 days long, instead of 365.* Present real calendar examples.

After completing the Example 1 estimate, return to the concept of a leap year. Ask, *Why is the Example 1 product an estimate?* **It does not include the days added by leap years.** Ask, *If Laura is nine years old, how many leap years has she probably lived through?* **two** *How do you know?* **Leap years happen every four years.** *If Laura lived through two leap years, how many more days old is she than 3,285?* **Two more days.**

### English Language Development Leveled Activities

| Emerging Level | Expanding Level | Bridging Level |
|---|---|---|
| **Developing Oral Language**<br>Write this problem in a place-value chart: 4,927 × 5 = ____. Give each student a copy. Point to the 7. Say, *Multiply the ones.* **Multiply the ones.** Then students proceed with this step. Point to the 2. Say, *Multiply the tens.* **Multiply the tens.** Then students proceed with this step. Point to the 9. Say, *Multiply the hundreds.* **Multiply the hundreds.** Then students proceed with this step. Point to the 4. Say, *Multiply the thousands.* **Multiply the thousands.** Then students proceed to multiply the thousands and add the regrouped hundreds. | **Do the Math**<br>Write a one-digit by four-digit multiplication problem on the board. Have volunteers come to the board one at a time to perform one of the 4 steps to solve the problem. Prior to completing his/her part of the problem, have each student identify the step they will be doing, using the following sentence frame: **I will be multiplying the number ____ in the (ones/ tens/hundreds/thousands) place with ____.** | **Number Game**<br>Write a multi-digit multiplication problem from the My Homework exercises. Distribute write-on/wipe-off boards. Have students race to find the solution. The first student to correctly solve the problem and explain each step of the solution gets to choose another My Homework exercise to write as the next problem. Have all winners assist other students by verbally talking them through the solution steps. |

## Teacher Notes:

NAME _____ DATE _____

# Lesson 9 Concept Web

## *Multiply by a Multi-Digit Number*

Use the concept web to identify each part of the partial product model
for 3 × 3,672.

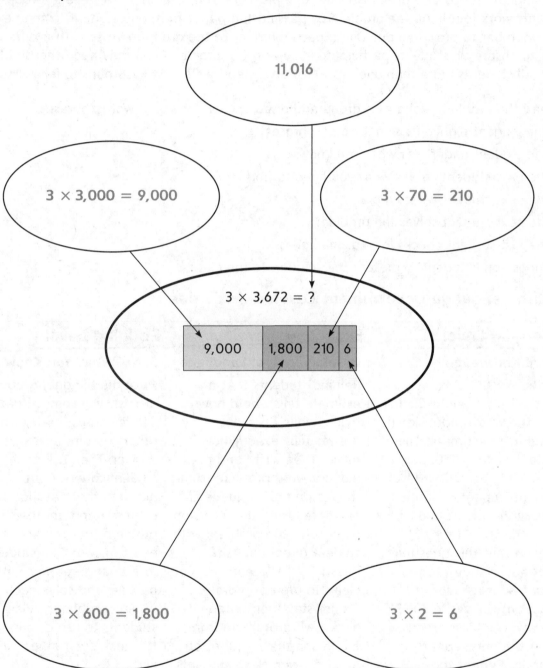

# Lesson 10 Problem-Solving Investigation
# Strategy: Estimate or Exact Answer
## English Learner Instructional Strategy

### Collaborative Support: Round the Table

Copy each word problem in the lesson on a separate sheet of paper. Place students into multilingual groups of 3 or 4. Distribute one problem from Exercises 1–8 to each group. Have students work jointly on the problem by passing the paper from one student to the next for each member to provide input. Direct each member of the group to write with a different color to ensure all students participate in solving the problem. You may need the problem to go to all students more than one time. Provide a step-by-step list for groups to follow, such as:

1. Read the problem aloud as a group and discuss. Look for signal words/phrases.

2. One student highlights signal words/phrases.

3. One student underlines what you know.

4. The next student circles what you need to find.

5. The next student writes a plan.

6. The next student solves the problem.

7. The last student checks for reasonableness.

8. Choose one student to present the solution to the class.

### English Language Development Leveled Activities

| Emerging Level | Expanding Level | Bridging Level |
|---|---|---|
| **Word Knowledge** | **Developing Oral Language** | **Show What You Know** |
| Write: 4 × 28. Ask, *4 × 28 is* about *how much?* Explain that the word *about* signals students to estimate. Then write 4 × 30 = 120. Say, *4 × 28 is* about *120.* Have students repeat chorally. Now write 3 × 21 and ask, *How much is 3 times 21?* Explain that this question does not include the word *about*, so it asks for an exact answer. Write 3 × 21 = 63. Say, *3 × 21 is exactly 63.* Have students repeat chorally. Repeat the exercise with additional multiplication problems. | Remind students that an estimate tells *about* how much, while an exact answer tells *exactly* how much. Then write and read aloud a word problem from the lesson that requires an estimate. Using the four-step plan, solve the problem with student direction. Ask questions if students are not clear, to prompt more proper statements. Repeat with a word problem from the lesson that requires an exact answer. Have students compare the two word problems. | Have multilingual groups discuss why some situations call for an estimate, while others require an exact answer. Then ask each student to write one problem that calls for an estimate and another problem that requires an exact answer. Tell students to exchange papers with a partner and solve each other's problems. Allow students to use calculators to check each other's work. |

### Teacher Notes:

NAME _____ DATE _____

# Lesson 10 Problem-Solving Investigation
## STRATEGY: Estimate or Exact Answer

Determine if an estimate or an exact answer is needed to solve each problem.

1. An office needs to buy **6 computers** and **6 printers**.
   Each **computer** costs **$384**. Each **printer** costs **$88**.
   **About** $2,400 will be spent on **computers**.
   What is the question?

| Understand | | | Solve |
|---|---|---|---|
| | | | The question is: |

| Item | Rounded Price (each) | Amount for 6 |
|---|---|---|
| Computer | 400 | 2,400 |
| Printer | 90 | 540 |

**Plan**

**Check**

2. **Each** fourth grade **class** reads a total of **495 minutes each** week.
   Suppose there are **4** fourth grade **classes**.
   How **many** minutes are read **each** week?

**Understand**

I know:

I need to find:

**Solve**

**Plan**
Facts that are important to solve the
problem are:

Words that indicate if an exact
answer **or** an estimate is needed are:

**Check**

Grade 4 • **Chapter 4** Multiply with One-Digit Numbers    **37**

# Lesson 11 Multiply Across Zeros
## English Learner Instructional Strategy

### Sensory Support: Physical Activities

Before the lesson, review the Distributive Property and the Zero Property of Multiplication. Invite students to share any mnemonic devices or pictures that help them remember these rules.

Before the lesson, review Examples 1 and 2 with students using the online Spanish Student Edition. For non-Spanish speaking ELs, provide an appropriate translation tool for the aide to utilize.

Use photos/realia to review English terms from the Problem Solving exercises, which may be unfamiliar, such as: *braces, trees, pool equipment, donate, charity,* and *money.*

### English Language Development Leveled Activities

| Emerging Level | Expanding Level | Bridging Level |
|---|---|---|
| **Background Knowledge** Write this problem in vertical format: $3,107 \times 2 =$ _____. Say, *First, I will estimate,* and then write: $3,000 \times 2 = 6,000$. Next, say, *Now I will solve the problem,* and then write: $3,107 \times 2 = 6,214$. Finally, explain that the last step is to check the answer. Point to 6,000 and say, *This is the estimate.* Have students repeat chorally. Then point to 6,214 and say, *This is the exact answer.* Have students repeat chorally. Plot both numbers on a number line. Gesture to the plotted numbers and say, *The exact answer is close to the estimate.* Prompt students to repeat. | **Developing Oral Language** Prepare a set of index cards that includes one card for each pair of students. On each card, write a problem that requires students to solve by multiplying across zeros. Pair students, and distribute one card to each pair. Have one partner estimate the answer and the other partner find the exact answer. Then have partners compare their results to see whether the estimated answer is reasonable. Have each pair report their finding back to you. | **Public Speaking Norms** Have each student write a one-digit by four-digit multiplication problem in which the four-digit number contains at least one zero. Tell students to exchange papers with a partner and solve each other's problem. Have students check their answers using a calculator. Then ask each student to present his or her solution to a group of emerging/ expanding students, explaining each step in solving the problem. |

### Teacher Notes:

NAME _____ DATE _____

# Lesson 11 Vocabulary Chart
## *Multiply Across Zeros*

Use the three-column chart to organize the vocabulary in this lesson.
Write the word in Spanish. Then write the correct terms to complete
each definition.

| English | Spanish | Definition |
|---|---|---|
| **estimate** | estimar | A number __close__ to an exact value. An estimate indicates __about__ how much. |
| **multiply** | multiplicar | An __operation__ on two numbers to find their __product__. It can be thought of as repeated __addition__. |
| **partial products** | productos parciales | A multiplication method in which the __products__ of each place value are found separately, and then __added__ together. |
| **Distributive Property** | propiedad distributiva | To multiple a __sum__ by a number, __multiply__ each addend by the number and __add__ the products. |

# Chapter 5 Multiply with Two-Digit Numbers

## What's the Math in This Chapter?

**Mathematical Practice 4: Model with mathematics.**

Write then read aloud this problem. *12 vans are going on a field trip. Each van is going to have 13 people in it. How many people are going on the field trip?*

Allow students time to think about the best way to solve the problem. Ask, *What operation should I use to solve?* **multiplication** If students respond **addition**, remind them that multiplication is repeated addition. However, adding 12 plus 13 will not tell them how many people total are going on the field trip.

First model the problem by drawing 12 rectangles and placing 13 X's in each rectangle. Then model using the base-ten blocks in Virtual Manipulatives. Have students use the models to determine the product: **156** Discuss which model they prefer to use. Say, When we model with mathematics, we use the math we know to help us solve real-world problems.

Display a chart with Mathematical Practice 4. Restate Mathematical Practice 4 and have students assist in rewriting it as an "I can" statement, for example: **I can use math models to solve real-world problems.** Post the new "I can" statement.

## Inquiry of the Essential Question:

### How can I multiply by a two-digit number?

Inquiry Activity Target: **Students come to a conclusion that they can use multiplication models to solve real-world problems.**

As an introduction to the chapter, present the Essential Question to students. The inquiry graphic organizer will offer opportunities for students to observe, make inferences, and apply prior knowledge of math models and multiplication representing the Essential Question. As they investigate, encourage students to draw, write, and collaborate with peers to demonstrate their observations and thinking. Then have students present additional questions they may have to a peer to extend discussions.

Regroup students and restate Mathematical Practice 4 and the Essential Question. Pose questions to reflect on what has been learned to guide students in making connections between the Mathematical Practice and the Essential Question.

NAME _____ DATE _____

# Chapter 5 Multiply with Two-Digit Numbers

## Inquiry of the Essential Question:

**How can I multiply by a two-digit number?**

**Read the Essential Question. Describe your observations (I see...), inferences (I think...), and prior knowledge (I know...) of each math example. Write additional questions you have below. Then share your ideas and questions with a classmate.**

| | | |
|---|---|---|
| $23 \times 40 = 23 \times (4 \times 10)$ | Write 40 as $4 \times 10$. | I see ... |
| $= (23 \times 4) \times 10$ | Associative Property of Multiplication | |
| | | I think... |
| $= 92 \times 10$ | Multiply. | |
| $= 920$ | Use mental math. | I know... |

10 + 6

34 { | $34 \times 10 = 340$ | $34 \times 6 = 204$ |

I see ...

I think...

$34 \times 16 = (34 \times 10) + (34 \times 6)$

$= 340 + 204$

$= 544$

I know...

| 56 | rounds to → | 60 | I see ... |
| $\times 37$ | rounds to → | $\times 40$ | |
| | | 2,400 | I think... |

I know...

Questions I have...

_____

_____

_____

# Lesson 1 Multiply by Tens
## English Learner Instructional Strategy

### Language Structure Support: Words in Context

Before teaching the lesson, explain how *think of* is used in Example 1, under "One Way: Use properties." Write and say these context sentences: *1. Think of me whenever there is a pretty sunset. 2. Think of three new uses for this object. 3. Think of my house as your house.* Discuss how, in Sentence 1, *think of* means "remember," and in Sentence 2, it means "brainstorm." Then underline *Think of* and *as* in Sentence 3, and write this sentence frame: *Think of ____ as ____.* Say, *When think of is followed by as, it means "consider to be the same." For example, another way to say Sentence 3 is, "Consider my house to be the same as your house."* Explain this is how *think of* is used in Example 1.

For oral language practice, ask students to share their answers to Problem Solving Exercises 24–25, using this conditional sentence frame: **If ____, then ____.** Have students add the sentence frame to their math journals.

### English Language Development Leveled Activities

| Emerging Level | Expanding Level | Bridging Level |
|---|---|---|
| **Word Knowledge** | **Listen and Write** | **Synthesis** |
| Stand next to a door. Say, *I am by the door.* Then move through the classroom, standing by different students. Say, *I am by (student's name).* Have students repeat chorally. Explain that, in this context, *by* means "beside." Then explain that *by* has a different meaning when used to describe multiplication. Write this on the board: $3 \times 10 = 30$. Say, *Three multiplied by ten equals thirty,* pointing to the multiplication sign as you say *multiplied by*. Have students repeat chorally. | Have students partner with a neighbor. Then have each student describe where they sit in relation to his or her partner using this sentence frame: **I sit by (student's name).** Next, have each pair write a multiplication sentence. Have them use this sentence frame to read aloud the sentence: **____ multiplied by ____ equals ____.** Finally, discuss with students how the uses of *by* differ in each sentence frame. | Write $3 \times 10 = 30$, and have students read it aloud. Then write $3 \times 100 = 300$, and have students read it aloud. Discuss the pattern of adding zero to the end of a product when multiplying by tens. Then, in multilingual groups, have students create and solve multiplication problems modeled after the example. |

**Teacher Notes:**

NAME _____ DATE _____

# Lesson 1 Concept Web
## *Multiply by Tens*

Use the concept web to write examples of multiples of 10.

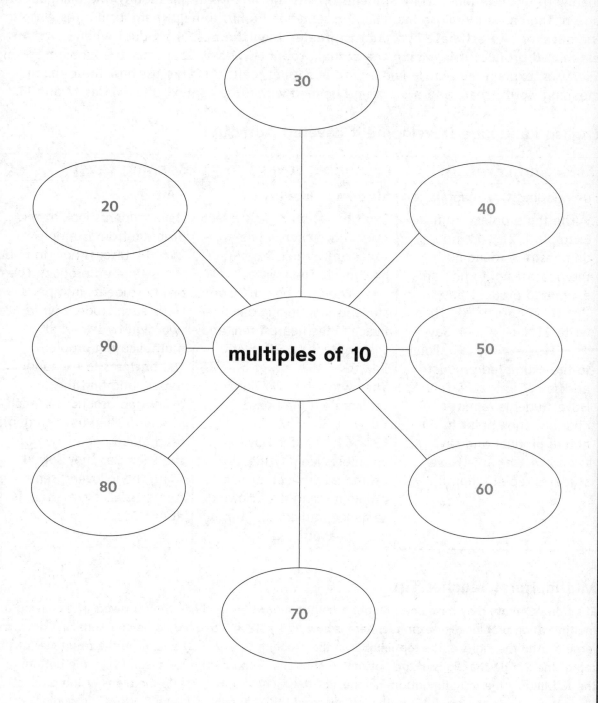

# Lesson 2 Estimate Products

## English Learner Instructional Strategy

### Vocabulary Support: Sentence Frames/Realia

Before the lesson, review the definitions for *round up* and *round down*. Invite students to share any pictures, gestures, or mnemonic devices they use to help them remember the meanings of these terms. Have students discuss the Talk Math question with a bilingual peer or aid. Then have them use the following sentence frames to help them form a response to the question: **An estimated product is greater than the actual product when _____. An estimated product is less than the actual product when _____.** Before the Problem Solving Exercises, explain the phrase *laid end to end* using realia. Also discuss how these multiple-meaning words; *span*, and *average* are defined within the context of Exercises 17 and 18.

### English Language Development Leveled Activities

| Emerging Level | Expanding Level | Bridging Level |
|---|---|---|
| **Developing Oral Language** | **Listen and Identify** | **Synthesis** |
| Write the problem from Example 1, 21 × 14, and demonstrate rounding to the nearest ten to find an estimated product. Say, *21 × 14 is **about** 200.* Then write 21 × 14 = 294. Say, *21 × 14 is **exactly** 294.* Point to the estimated product and say, *This is an estimate.* Have students repeat chorally. Then point to the actual product and say, *This is an exact answer.* Have students repeat chorally. | Explain that an *estimate* tells *about* how much, while an *exact answer* tells *exactly* how much. Then write 33 × 19 on the board. Guide students in rounding each factor to the nearest ten, and then write 30 × 20 = 600. Say, *33 × 19 is about 600.* Then write the following problems on the board: 88 × 21; 36 × 37; 72 × 49; 53 × 64; 12 × 58. Have volunteers identify the estimated product for each problem, using the following sentence frame: _____ times _____ is about _____. | Have students look at this problem from Example 1: 21 × 14. Discuss how to find the estimated product. Then divide students into pairs. Have each student write five two-digit by two-digit multiplication problems. Tell partners to exchange papers and find the estimated product for each problem. Then have students read aloud one or two of the exercises they solved using the following sentence frame: _____ times _____ is about _____. |

### Multicultural Teacher Tip

ELs from Vietnam may have been taught a unique algorithm to check their answers after solving a multiplication problem. For example, here is how 473 × 12 = 5,676 would be checked: First draw a large X. Add the digits of the top number in the problem (4 + 7 + 3) and write the result (14) at the top of the X. Add the digits of the bottom number (1 + 2) and write the result (3) at the bottom of the X. Multiply the top and bottom numbers of the X (14 × 3 = 42), add the digits of the product (4 + 2) and write the result (6) in the left space of the X. Finally, add the digits of the original answer (5 + 6 + 7 + 6 = 24), add the digits of the result (2 + 4), and write the number (6) in the right space of the X. If the numbers to the left and right of the X match, the answer is correct (6 = 6, so 5,676 is the correct answer). Ask the student to model this algorithm for the class.

NAME _____ DATE _____

# Lesson 2 Guided Writing

## Estimate Products

**How do you estimate products?**

Use the exercises below to help you build on answering the Essential Question. Write the correct word or phrase on the lines provided.

**1.** Rewrite the question in your own words.

See students' work.

_____

**2.** What key words do you see in the question?

estimate, products

**3.** When you __round__ a number, you make it easier to work with by changing the __value__ of the number.

**4.** To __round__ a number to the nearest ten, look at the __ones__ place.

**5.** If the value in the __ones__ place is greater than 5, round __up__ to the nearest ten.

**6.** If the value in the ones place is less than 5, round __down__ to the __nearest__ ten.

**7.** An __estimate__ is a number close to an exact value. An estimate indicates __about__ how much.

**8.** How do you estimate products?

Round each factor to the nearest ten, then multiply to find the estimated product.

# Lesson 3 Inquiry/Hands On: Use the Distributive Property to Multiply

## English Learner Instructional Strategy

### Vocabulary Support: Activate Prior Knowledge

Before the lesson, review *Distributive Property* and its cognate *propiedad distributiva*. Invite students to share any mnemonic devices or pictures that help them remember this rule.

Then review how to use the Distributive Property to multiply a one-digit number by a two-digit number. Discuss the example that appears before the Draw It activities on Student Edition page 293. Write: $3 \times 11$. Circle the 11 and then draw a model of it using all ones blocks. Ask, *Can we regroup?* **yes** *How many tens are in 11?* **one** Draw a new model for 11 that uses one tens blocks and one ones block. Then write $3 \times 11 = 3 \times (10 + 1)$ and have students use these sentence frames to help solve the problem: _____ **times** _____ **tens equals part of the product:** _____. _____ **times** _____ **ones equals part of the product:** _____. _____ **+** _____ **= the whole product:** _____. Finally, tell students the Distributive Property can help them multiply 2 two-digit numbers.

### English Language Development Leveled Activities

| Emerging Level | Expanding Level | Bridging Level |
|---|---|---|
| **Listen and Identify** | **Share What You Know** | **Communication Guide** |
| Write: *dimensions.* Say the word and have students chorally repeat. Then explain that the dimensions of an area model are its *length* and its *width.* Write: *length/long, width/wide.* Then have students look at the area model at the bottom of Student Edition page 293. Trace along its first column with your pointer finger. Say, *This is the length. How many units?* **12** Then say, *The length of this model is 12 units. It is 12 units long.* Next, trace along the model's first row. Say, *This is the width. How many units?* **15** Then say, *The width of this model is 15 units. It is 15 units wide.* | Have students do the Emerging-Level activity. Then assign the Practice It Exercises. Do Exercise 3 as a class, taking care to discuss how the area model was created. Then pair up students to do Practice It Exercises 4-6, assigning each set of partners one item to work on together. Have students discuss how to set up their models using these sentence frames: **The model is _____ units long. The model is _____ units wide. The width has _____ tens and _____ ones.** Have partners solve then present their model and solution to the group. | Pair students. Have one do Talk About It Exercise 1 and the other do Exercise 2, using these sentence frames to help with forming a response: **Regroup _____ as _____ ones + _____ tens. Rewrite the equation as: _____ × (_____ + _____). Next, multiply _____ by _____ tens to find one part of the product: _____. Then multiply _____ by _____ ones to find the other part of the product: _____. Finally, add the partial products to find the whole product: _____ + _____ = _____.** Then have pairs explain their solutions to each other. |

NAME _____ DATE _____

# Lesson 3 Vocabulary Definition Map
*Inquiry/Hands On: Use the Distributive Property to Multiply*

Use the definition map to write a description and list characteristics about the vocabulary word or phrase. Write or draw math examples. Share your examples with a classmate.

**My Math Vocabulary:**

**Distributive Property of Multiplication**

**Characteristics from Lesson:**

A two-digit number can be decomposed into the sum of the values of the __tens__ place and the __ones__ place.

The __parentheses__ let you know that you should do that operation first.

$4 \times 12 = 4 \times (\underline{10} + \underline{2})$
$= (4 \times \underline{10}) + (4 \times \underline{2})$
$= \underline{40} + \underline{8}$
$= \underline{48}$

**Description from Glossary:**

To multiply a sum by a number, multiply each addend by the number and add the products.

**My Math Examples:**
See students' examples.

# Lesson 4 Multiply by a Two-Digit Number

## English Learner Instructional Strategy

### Sensory Support: Diagrams/Models

During Math in My World, Example 1, ensure students understand how an area model is used when multiplying two-digit numbers. Refer students to the area model in Chapter 4, Lesson 5, on page 223. Review how the numbers in the model are derived from $2 \times 24$. Then have students compare the Chapter 4 area model to the one in this lesson, on page 299. Point out that the Chapter 4 area model has one row because the one factor in its corresponding multiplication problem has one digit. Contrast this with the Chapter 5 area model, which has two rows because both factors in its corresponding multiplication problem have two digits. Then explain how the numbers in the Chapter 5 area model were derived from $27 \times 12$. Write: $(20 + 7) \times (10 + 2)$. Then have students help you solve the problem. Ask, *What is 20 × 10?* **200** *What is 20 × 2?* **40** *What is 7 × 10?* **70** *What is 7 × 2?* **14** Show students how these products correspond to the area model. Then solve the problem by adding the partial products. **324**

### English Language Development Leveled Activities

| Emerging Level | Expanding Level | Bridging Level |
|---|---|---|
| **Developing Oral Language** | **Word Recognition** | **Anchor Charts** |
| Draw an area model on the board. Label it using this multiplication problem from Example 1: 27 × 12. Then point to each factor and say, *This is a factor*. Next point to each partial product and say, *This is a partial product*. Finally, randomly point to various numbers, identifying each as a factor or a partial product. Have students show thumbs-up, if you label a number correctly, or thumbs-down, if you label a number incorrectly. | Draw an area model on the board. Label it using the multiplication problem from Example 1: 27 × 12. Randomly point to various numbers, identifying each as either a *factor* or a *partial product*. Then have students identify each using the following sentence frames: _____ **is a partial product;** _____ **is a factor.** Then have students create an area model for 36 × 26. | Have multilingual groups create anchor charts using partial products or the standard algorithm to solve two-digit multiplication problems. Assign students problems from the lessons. Have students label all chapter vocabulary. Direct students to show the sequence of steps for using partial products or the steps to work through the standard algorithm to solve their assigned multiplication problem. Post charts and discuss as a class. |

**Teacher Notes:**

NAME _____ DATE _____

# Lesson 4 Vocabulary Cognates
*Multiply by a Two-Digit Number*

Use the Glossary to define the math word in English and in Spanish in the word boxes. Write a sentence using your math word.

| **product** | **producto** |
|---|---|
| **Definition** | **Definición** |
| The answer or result of a *multiplication* problem. It also refers to expressing a number as the *product* of its *factors*. | Respuesta o resultado de un problema de *multiplicación*. Además, un número puede expresarse como el *producto* de sus *factores*. |

**My math word sentence:**
Sample answer: The product of 2 and 3 is 6.

| **partial products** | **productos parciales** |
|---|---|
| **Definition** | **Definición** |
| A multiplication method in which the *products* of each *place value* are found separately, and then added together. | Método de multiplicación por el cual los *productos* de cada *valor posicional* se hallan por separado y luego se suman entre sí. |

**My math word sentence:**
Sample answer: The partial products of 4 × 12 are 4 × 10 = 40 and 4 × 2 = 8.

# Lesson 5 Solve Multi-Step Word Problems

## English Learner Instructional Strategy

### Sensory Support: Illustrations/Drawings

Before the lesson, write *operation* on the board. Tell students that this is a multiple-meaning word that can mean "a medical procedure," "a planned activity" or "a business," among other things. Ensure students understand that in a math context it means "a mathematical process, such as addition, subtraction, multiplication, or division." Provide illustrations and concrete examples to support understanding.

Also discuss the term real-world. Say, *In this lesson, we will write and solve real-world math problems.* Explain that *real-world* is an adjective used to describe things that happen in the "real world," or in people's everyday lives. Invite students to offer *real-world* examples of times they use math.

Encourage students to draw a picture or use models to help them visualize Exercise 8.

## English Language Development Leveled Activities

| Emerging Level | Expanding Level | Bridging Level |
|---|---|---|
| **Listen and Identify** | **Act It Out** | **Signal Words/Phrases** |
| Write the symbol for each of the four operations $(+, -, \times, \div)$. Then point to each symbol and say the operation it represents (*add, subtract, multiply, divide,* respectively). Finally, randomly point to the symbols, and have students say the operation each symbol represents. Write $(3 \times 62) - 25 + n = 166$. Point to the parentheses and say, *We do this operation first. What is it?* **multiply** Have students multiply. (186) Then point to each following symbol, have students identify then do the operation. The unknown is 5. | Write the operation words: *add, subtract, multiply*, and *divide*. Below each word, write its corresponding operation symbol $(+, -, \times,$ and $\div$, respectively). Then create and recite simple word problems for students to act out. For example, say, *There are 9 students. They divided into three equal groups. How many students are in each group?* Have students act out the problem and then identify the operation used to find the answer. **division; 3** | Review with students the terms: *add, subtract, multiply*, and *divide* and their corresponding operation symbols $(+, -, \times,$ and $\div)$. Then have students create a graphic organizer listing terms associated with each of the four operations. For example, for *subtract*, students might list: **difference, less than, fewer.** Finally, have students write a real-world word problem using at least one of the signal words/phrases from their graphic organizer. |

**Teacher Notes:**

NAME _____ DATE _____

# Lesson 5 Note Taking
## *Solve Multi-Step Word Problems*

Read the question. Write words you need help with and research each word. Use your lesson to write your Cornell notes. Write or draw math examples to explain your thinking. Share your examples with a classmate.

**Building on the Essential Question**

How do you solve multi-step word problems?

**Words I need help with:**

See students' words.

**Notes:**

An operation is a mathematical process such as __addition__ (+), __subtraction__ (−), __multiplication__ (×), or __division__ (÷).

A __variable__ is a letter or symbol used to represent an __unknown__ quantity.

__Parentheses__ ( ) are the enclosing symbols which indicate that the terms within are a unit.

When you see the words: *less than, fewer, remains,* or *difference* in a word problem, the operation to use is most likely __subtraction__.

When you see the words: *total, altogether, both,* or *sum* in a word problem, the operation to use is most likely __addition__.

When you see the words: *times, each, every day, at this rate,* or *product* in a word problem, the operation to use is most likely __multiplication__.

**My Math Examples:**

See students' examples.

# Lesson 6 Problem-Solving Investigation Strategy: Make a Table

## English Learner Instructional Strategy

### Collaborative Support: Partners Work/Pairs Check

Write table/tabla on a Spanish cognate chart. Introduce the word, and provide a math example. Utilize other appropriate translation tools for non-Spanish speaking ELs. Discuss alternative meanings for the word, such as "a graphic organizer," "a piece of furniture," and "to pause until a later time." Provide photos and concrete examples to support understanding. Ensure students understand that, in the context of this lesson, table is a type of graphic organizer.

Divide students into pairs to solve Apply the Strategy Exercises 1–4. For the first problem, have one student complete the problem while the other student coaches. Then, for the second problem, have partners switch roles. Once they have completed both problems, tell students to work with another pair to check answers. When both pairs have agreed on the answers, have them shake hands and continue working with their original partners on the next two problems.

## English Language Development Leveled Activities

| Emerging Level | Expanding Level | Bridging Level |
|---|---|---|
| **Word Knowledge**<br>Stand by a table and point to it. Say, *This is a table.* Then explain that *table* is also a name for a type of chart. Draw the following table: | **Building Oral Language**<br>Draw the following table: | **Number Game**<br>Draw a blank table that has 2 rows and 5 columns. Use this to review the terms *table*, *row*, and *column*. Then have students think of a number pattern, such as: **add 8, subtract 3, or multiply by 4.** Ask students to keep their number patterns secret. Then have them copy the table you drew, filling it in using the number pattern they chose, but leaving three sections of the table blank. Have students exchange papers with a partner and complete each other's table. |

Emerging Level table:

| Number of Dogs | Number of Legs |
|---|---|
| 1 | 4 |
| 2 | 8 |
| 3 | 12 |

Say, *This is a table.* Explain that this kind of table can be used to solve problems. Identify and discuss the number pattern in this table. Visually support with pictures of dogs.

Expanding Level table:

| 1 book | $7 |
|---|---|
| 2 books | $14 |
| 3 books | $21 |

Say, *This is a table.* Point out that 1 book costs $7, 2 books cost $14, and so on. Guide students in recognizing the pattern, "multiply by 7." Then ask students to extend the table to include the cost of 4 books. Repeat with greater numbers of books.

NAME _____ DATE _____

# Lesson 6 Problem-Solving Investigation
## STRATEGY: Make a Table

album

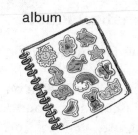

Make a table to solve each problem.

1. **A page** from **Dana's** album is shown.
   Dana puts the **same** number of stickers on **each** page.
   **She** (Dana) has **30 pages** of stickers.
   **How many** stickers does she have **in all**?

| Understand | Solve |
|---|---|
| I know: | |
| | | Pages | 1 | 10 | 20 | 30 | |
| | | Stickers | 12 | 120 | 240 | 360 | |
| I need to find: | Dana has __360__ stickers. |
| **Plan** | **Check** |
| I will make a __table__ to solve. | |

2. West Glenn School has **23 students** in **each** class.
   There are **6** fourth grade **classes**.
   **About** **how many** fourth grade **students** are there **in all**?

| Understand | Solve |
|---|---|
| I know: | |
| | | Classes | 1 | 2 | 3 | 4 | 5 | 6 | |
| | | Students | 20 | 40 | 60 | 80 | 100 | 120 | |
| I need to find: | There are **about** __120__ students. |
| **Plan** | **Check** |
| I will make a __table__ to solve. | |

Grade 4 • **Chapter 5** Multiply with Two-Digit Numbers  **45**

# Chapter 6 Divide by a One-Digit Number

## What's the Math in This Chapter?

### Mathematical Practice 5: Use appropriate tools strategically.

Write this incorrect division sentence on the board: $2\overline{)39}$ with $41$ above. Say, *Look at this division problem. Do you notice anything wrong or incorrect?* Give students time to observe on their own. Then have them turn and talk with a peer. Discuss as a group. Students should notice that the quotient is incorrect. Ask, *How does division affect numbers?* Guide students to observe that the quotient should always be less than the dividend.

Discuss that if we use estimation with compatible numbers as a tool before we solve, we can determine if our answer is close. Write $40 \div 2 = ?$ and have students use mental math to solve. **20** Say, *So, we know then that our quotient should be close to 20. 41 is not close to 20 and 41 is more than the dividend 39. We can use estimation and what we know about math as tools to help us understand division problems.* Have students solve. **19 R1** Say, *19 R1 is close to 20. Our solution is reasonable.*

Display a chart with Mathematical Practice 5. Restate Mathematical Practice 5 and have students assist in rewriting it as an "I can" statement, for example: **I can use math tools to better understand division.** Have students draw or write examples of using math vocabulary precisely/clearly. Post the chart in the classroom.

## Inquiry of the Essential Question:

### How does division affect numbers?

Inquiry Activity Target: **Students come to a conclusion that division consists of other operations.**

As an introduction to the chapter, present the Essential Question to students. The inquiry graphic organizer will offer opportunities for students to observe, make inferences, and apply prior knowledge of operations representing the Essential Question. As they investigate, encourage students to draw, write, and collaborate with peers to demonstrate their observations and thinking. Then have students present additional questions they may have to a peer to extend discussions.

Regroup students and restate Mathematical Practice 5 and the Essential Question. Pose questions to reflect on what has been learned to guide students in making connections between the Mathematical Practice and the Essential Question.

NAME _____ DATE _____

# Chapter 6 Divide by a One-Digit Number

## *Inquiry of the Essential Question:*

**How does division affect numbers?**

Read the Essential Question. Describe your observations (I see…), inferences (I think…), and prior knowledge (I know…) of each math example. Write additional questions you have below. Then share your ideas and questions with a classmate.

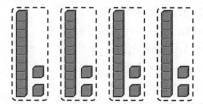

There is 1 ten and 2 ones, or 12 in each group, so 48 ÷ 4 = 12.

I see …

I think…

I know…

---

$$\begin{array}{r} 5\,\text{R}1 \\ 5{\overline{\smash{)}26}} \\ -25 \\ \hline 1 \end{array}$$

Divide
Multiply
Subtract

I see …

I think…

I know…

---

| | |
|---|---|
| 36 ÷ 6 = 6 | 6 × 6 = 36 |
| 360 ÷ 6 = 60 | 6 × 60 = 360 |
| 3,600 ÷ 6 = ? | 6 × 600 = 3,600 |

So, 3,600 ÷ 6 = 600

I see …

I think…

I know…

---

Questions I have…

_____

_____

_____

# Lesson 1 Divide Multiples of 10, 100, and 1,000

## English Learner Instructional Strategy

### Language Structure Support: Words in Context

Before the lesson, write *multiples* and *dividend* and their Spanish cognates, *múltiplos* and *dividendo* on a cognate chart. Introduce each word, and provide math examples. Utilize other appropriate translation tools for non-Spanish speaking ELs.

Read aloud the paragraph describing the Migration Table, for Problem Solving Exercises 23–24. Point out the term *factors*. Invite students to tell you what a factor is, in a math context. **A *factor* is a number that divides a whole number evenly or a number that is multiplied by another number.** Then explain that factor is a multiple-meaning word. Briefly discuss that the -s suffix shows that factor is plural. Then model using context clues to help you define *factors* as "causes or reasons for something happening." Ask, *What are the **factors** in this paragraph?* **climate and food availability** *What do they cause?* **animal migration**

### English Language Development Leveled Activities

| Emerging Level | Expanding Level | Bridging Level |
|---|---|---|
| **Listen and Identify**<br>Write: **1.** $16 \div 2 = 8$; **2.** $160 \div 2 = 80$; **3.** $1,600 \div 2 = 800$. In Item 1, point to the 16 and then the 8, and say, *There is no zero.* Then in Item 2, point to the 160 and then the 80, and say, *There is one zero. The factors have 1 zero and the product has 1 zero.* Repeat with Item 3, identifying the pattern of 2 zeros. Repeat the process with: **4.** $49 \div 7 = 7$; **5.** $490 \div 7 = 70$; **6.** $4,900 \div 7 = 700$. Help students identify the zero patterns, using this sentence frame: **The dividend has _____ zero(s) and the quotient has _____ zero(s).** | **Show What You Know**<br>Write: **1.** $21 \div 7 = 3$; **2.** $210 \div 7 = 30$; **3.** $2,100 \div 7 = 300$ Discuss the equations. Point to the multiples of 21 and say, *These are **dividends**.* Point to the multiples of 3 and say, *These are **quotients**.* Ask, *What pattern do you see in these three equations?* Guide students in recognizing that the number of zeros in each *dividend* is the same as the number of zeros in each *quotient*. Have small groups use the same pattern to find the unknowns in these equations: **4.** $54 \div 9 = m$; **5.** $540 \div 9 = n$; **6.** $5,400 \div 9 = p$. | **Developing Oral Language**<br>Write: **1.** $32 \div 8 = 4$; **2.** $320 \div 8 = 40$; **3.** $3,200 \div 8 = 400$. Ask a volunteer to read Item 1. Explain that this equation is the *basic fact*. Then have students use the following sentence frame to describe the next two equations: **When the dividend has _____ zeros, the quotient has _____ zeros.** Have multilingual groups write other basic facts and patterns like the one in the example. Have them use the sentence frame above to describe their equations that contain multiples of 10. |

**Teacher Notes:**

NAME _____ DATE _____

# Lesson 1 Vocabulary Cognates
## *Divide Multiples of 10, 100, and 1,000*

Use the Glossary to define the math word in English and in Spanish in the word boxes. Write a sentence using your math word.

| **dividend** | **dividendo** |
|---|---|
| **Definition** | **Definición** |
| A number that is being *divided*. | Número que se divide. |
| **My math word sentence:** ||
| Sample answer: In the division sentence 6 ÷ 2 = 3, 6 is the dividend. ||

| **multiple** | **múltiplo** |
|---|---|
| **Definition** | **Definición** |
| A multiple of a number is the *product* of that number and any whole number. | Un múltiplo de un número es el *producto* de ese número y cualquier otro número natural. |
| **My math word sentence:** ||
| Sample answer: The numbers 4, 6, 8, 10, and 12 are all multiples of 2. ||

# Lesson 2 Estimate Quotients
## English Learner Instructional Strategy

### Sensory Support: Illustrations, Diagrams, Drawings

Before the lesson, review the terms *compatible numbers, fact family, basic fact,* and *place value*. Write the terms on chart paper, and ask students to generate a definition and math example for you to write next to each term. Also invite students to suggest any illustrations, diagrams, or drawings that help them remember the meanings of the terms, and note them beside the terms' definitions. Encourage students to copy this information in their math journals.

To help students discuss the Talk Math problem, provide this communication guide:

**To estimate $4,782 ÷ 6, first round $4,782 up to _____.**
**Then divide _____ by 6.**
**The basic fact related to the numbers in this equation is**
**_____ × _____ = _____. The estimate is _____.**

### English Language Development Leveled Activities

| Emerging Level | Expanding Level | Bridging Level |
|---|---|---|
| **Academic Vocabulary** | **Act It Out** | **Think-Pair-Share** |
| Have two students cooperatively perform a simple task, such as lifting and moving a chair. Then say, *[Student 1] and [Student 2] work well together. They are compatible.* Next, write and say: *compatible numbers.* Have students repeat chorally. Explain that compatible numbers are numbers that work well together. Discuss how compatible numbers can be used to estimate showing examples from the lesson. | Hold up 13 pencils, and then call 3 students to the front of the room. Say, *I have 13 pencils. **About** how many will each person get?* Remind students that the word *about* shows they should estimate. Then write 13 ÷ 3 and say, *13 and 3 are **not** compatible numbers; they do **not** work well together.* Replace 13 with 12, and say, *12 and 3 **are** compatible numbers. They work well together because 12 divided by 3 results in a whole number, 4.* Distribute the pencils one at a time to the 3 students. Then say, *Our estimate was close. Each person has **about** 4 pencils.* | Have students compare two strategies they have used for estimating: rounding and using compatible numbers. Then ask students which estimating strategy they think is better. Have them come up with two reasons their chosen method is better for them. Then ask students to meet with a partner to discuss their ideas. Finally, have students share their preferred method and supporting reason with the group. |

**Teacher Notes:**

NAME _____ DATE _____

# Lesson 2 Vocabulary Definition Map
## *Estimate Quotients*

Use the definition map to write a description and list characteristics about the vocabulary word or phrase. Write or draw math examples. Share your examples with a classmate.

My Math Vocabulary:

**compatible numbers**

Characteristics from Lesson:

Compatible numbers can be used to <u>estimate</u> quotients.

Sometimes you round down to find the compatible number. The compatible numbers for estimating 453 ÷ 7 are <u>420</u> ÷ 7 = 60.

Sometimes you round up to find the compatible number. The compatible numbers for estimating 453 ÷ 8 are <u>480</u> ÷ 8 = 60.

Description from Glossary:

Numbers in a problem or related numbers that are easy to work with mentally.

My Math Examples:
See students' examples.

# Lesson 3 Inquiry/Hands On: Use Place Value to Divide

## English Learner Instructional Strategy

### Vocabulary Support: Make Connections

Before the lesson, write and say *remainder*. Have students chorally repeat. Explain that *remainder* is a word with multiple meanings. Model that one of its meanings is "something that is left after other parts have been removed." Point out that a synonym for this meaning of *remainder* is *leftover*. Discuss how in English, food that is not eaten at a meal is commonly called *leftovers,* as in: *Today, we had turkey for dinner. Tomorrow we will eat the **leftovers**. We will use the **leftover** turkey to make sandwiches.* Explain that in this example, a *whole* turkey was served at dinner and *part* of the turkey was not eaten—*part* of the turkey is what was left. Invite volunteers to model use of the term *leftover* using this sentence frame: **Yesterday, we ate _____. Today, I will eat leftover _____.** Then explain how the non-math meaning of *remainder* relates to its math meaning. Say, *In math, the **remainder** is part of the dividend that was **left over***. Have students preview an example in the lesson where part of the dividend is left over.

## English Language Development Leveled Activities

| Emerging Level | Expanding Level | Bridging Level |
|---|---|---|
| **Listen and Identify** Review the term *equal groups*. Model dividing 12 counters into three groups of 4 counters. Say, *These are **equal** groups. Each group has the same number of counters. How many counters are in each group?* **four** Divide the 12 counters into two groups of 6 counters. Ask, *Are these equal groups?* **yes** *How many are in each group?* **six** Now, divide 7 counters into two groups, one with 3 counters and another with 4 counters. Ask, *Are these equal groups?* **no** Say, *I can make two equal groups of 3 counters. But, 1 counter is left over. It is the **remainder***. | **Academic Vocabulary** Write *Remainder,* and the equation 68 ÷ 5 = 13 R3 horizontally and in a division bracket. Point to each uppercase R in the equations. Say, *This is an abbreviation. An abbreviation is a short way to write a word.* Explain that some abbreviations use the first letter of a word. Point to the word *Remainder* and ask, *What do you think R means?* **remainder** Point out that in math some single-letter abbreviations are capitalized, while others are not. | **Exploring Language Structure** Do the Expanding-Level activity with students. Then return to this equation: 68 ÷ 5 = 13 R3. Tell students that one way to say it is: 68 divided by 5 equals 13. The remainder is 3. **However,** another way to say it is: 68 divided by 5 is 13 with a remainder of 3. Point out that in the second version, *is* takes the place of *equals* and *with a remainder of* takes the place of *The remainder is.* During the Apply It exercises, have students practice saying division equations with remainders using this sentence frame: **_____ divided by _____ is _____ with a remainder of _____.** |

**Teacher Notes:**

NAME _____ DATE _____

# Lesson 3 Concept Web

## *Inquiry/Hands On: Use Place Value to Divide*

Use the concept web to write each part of a division equation.

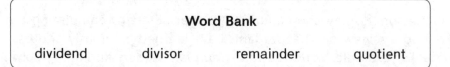

**Word Bank**

dividend          divisor          remainder          quotient

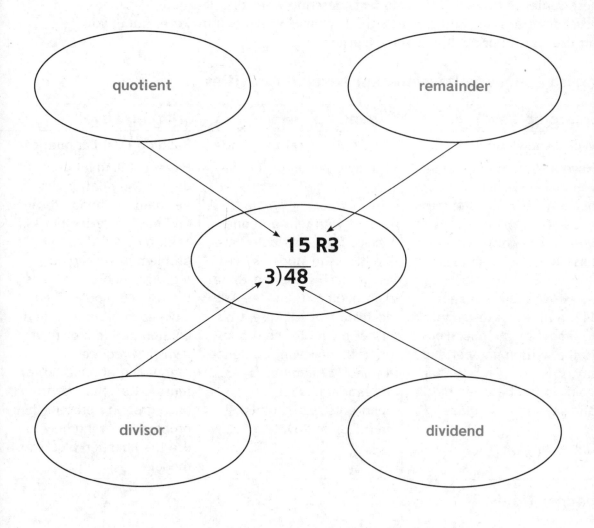

# Lesson 4 Problem-Solving Investigation Strategy: Make a Model

## English Learner Instructional Strategy

### Vocabulary Support: Build Background Knowledge

Write *model* and its Spanish cognate, *modelo*. Introduce the word, and provide math examples such as drawings, number lines and base-ten blocks.

Throughout the lesson, display and fill in a KWL chart. Start by completing the first column during a review of what students have learned previously about problem solving. For example, write the *four-step plan:* Understand, Plan, Solve, Check. Have students suggest previous strategies they have utilized such as: *make a table, act it out,* and *draw a picture.* Introduce the strategy *make a model* and write what students want to know. Following the lesson, display these sentence frames to help students describe what they learned, and then record their answers in the chart's third column: **When you make a model you can use _____. During the lesson I learned _____.**

### English Language Development Leveled Activities

| Emerging Level | Expanding Level | Bridging Level |
|---|---|---|
| **Word Knowledge** | **Developing Oral Language** | **Building Oral Language** |
| Have a student pose while you quickly sketch him or her. Say, *[Student's name] is a **model**. An artist draws a **model**.* Then display some base-ten blocks, and say, *These are another type of **model**.* Write the number 142, and use base-ten blocks to model the number. Point to the written number and say, *This is a number.* Then point to the base-ten blocks and say, *This is a **model**.* | Display a hundreds flat. Say, *This is a hundreds **model**. The model equals 1 hundred.* Repeat with tens and ones blocks. Then distribute base-ten blocks to students, and assign each student a three-digit number to model. Ask each student to present his or her model to the group, using the following sentence frames: **My number is _____. My model has _____ hundred(s), _____ ten(s), and _____ one(s).** | Have multilingual groups discuss the relationship between the hundreds, tens, and ones base-ten blocks. Then model a division problem in which you regroup hundreds or tens. Discuss the model. Then have each group create a division problem of their own that requires a regrouping of hundreds or tens. Ask a volunteer from each group to present their problem, showing how to use the base-ten blocks as a model. |

**Teacher Notes:**

NAME _____ DATE _____

# Lesson 4 Problem-Solving Investigation

## STRATEGY: Make a Model

**Make a model to help you solve each problem.**

1. **Casey's** <u>mom</u> is the baseball <u>coach</u> for his team.    baseball
   She (mom/coach) **spent $150** on baseballs.
   <u>Each</u> baseball **cost $5.**
   **How many** baseballs did **she buy?**

| Understand | Solve |
|---|---|
| I know: <br><br><br> I need to find: | |
| **Plan** <br> I will make a <u>model</u> to divide. <br> I will use base-ten <u>blocks</u> . | **Check** |

2. **Each** flowerpot costs **$7.**
   **How many** flowerpots can be bought **with $285?**

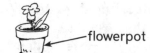 —flowerpot

| Understand | Solve |
|---|---|
| I know: <br><br><br> I need to find: | |
| **Plan** <br> I will make a model to <u>divide</u> . <br> I will use <u>base-ten</u> blocks. | **Check** |

# Lesson 5 Divide with Remainders

## English Learner Instructional Strategy

### Sensory Support: Mnemonics

Before the lesson, write *quotient* and its Spanish cognate, *cociente*. Introduce the word, and provide a math example. Utilize other appropriate translation tools for non-Spanish speaking ELs.

Write $15 \div 3 = 3\overline{)15}^{\,5}$. With students' help, label the *dividends*, *divisors*, and *quotients* for each division example.

Finally, review the My Vocabulary Card for *remainder*. Write the term and underline *remain* within it. Say, *The **remainder** is the number that **remains**. It is **leftover**.* Have students repeat chorally. Then refer students to the quotient in Math in My World, Example 1. Write it on the board: 19 R1. Then say, *The quotient is 19 with a **remainder** of 1.* As you say *remainder*, circle the R. Explain that *R* is an abbreviation.

## English Language Development Leveled Activities

| Emerging Level | Expanding Level | Bridging Level |
|---|---|---|
| **Listen and Identify** | **Developing Oral Language** | **Public Speaking Norms** |
| Write the division problem from Example 1 on the board, and solve it as shown in the lesson. Write the word *quotient* above the 19. Circle the 19 and say, *This is the quotient.* Have students repeat chorally. Then write remainder above the 1. Circle the 1 and say, *This is the remainder.* Have students repeat chorally. Finally, invite students to point to either the 19 or the 1 and correctly identify it as *quotient* or *remainder*. | Write the division problem from Example 1 on the board, and solve it as shown in the lesson. Then point to the divisor, dividend, quotient, and remainder randomly, having students name each of these components of the equation. Finally, assign each student a division problem from this lesson's Reteach worksheet. Have students describe the solution to their problem using the following sentence frames: **The quotient is _____. The remainder is _____.** | Have multilingual groups work together to check their answers to Independent Practice Exercises 3–11. Then ask a volunteer from each group to demonstrate how they solved the division problem using appropriate lesson vocabulary. Finally have another volunteer explain how they used multiplication and, if necessary, addition to check the answer. |

**Teacher Notes:**

NAME _____ DATE _____

# Lesson 5 Note Taking

## *Divide with Remainders*

Read the question. Write words you need help with and research each word. Use your lesson to write your Cornell notes. Write or draw math examples to explain your thinking. Share your examples with a classmate.

| **Building on the Essential Question** | **Notes:** |
|---|---|
| How can you divide with remainders? | Steps to __divide__ a two-digit number by a one-digit number. <br><br> 1. Divide the __tens__ <br> • multiply <br> • __subtract__ <br> • compare <br> • bring __down__ the ones <br><br> 2. Divide the __ones__ <br> • __multiply__ <br> • subtract <br> • __compare__ <br> • bring down the __ones__ |
| **Words I need help with:** <br> See students' words. | If there are no __numbers__ to bring down, the number left over is called the __remainder__. <br><br> $$\begin{array}{r} 19\,R1 \\ 2\overline{)39} \\ -2\!\downarrow \\ \hline 19 \\ -18 \\ \hline 1 \end{array}$$ ← number left over |
| **My Math Examples:** <br> See students' examples. | |

# Lesson 6 Interpret Remainders
## English Learner Instructional Strategy

### Vocabulary Support: Signal Words/Phrases

Discuss the multiple-meaning word *interpret*, which means "explain" within the context of this lesson. Then say, *In this lesson, you will interpret, or explain, what the remainders in word problems mean.*

Create a *Remainders* word web that includes signal words or phrases, such as *extra, additional,* and *left/left over.* Discuss the terms, taking extra time with *left,* which has multiple meanings. Write then say aloud the sentence, *How many pieces are left?* Say, *Another way to say this is, "How many pieces remain?"*

### English Language Development Leveled Activities

| Emerging Level | Expanding Level | Bridging Level |
|---|---|---|
| **Choral Responses** Write $10 \div 3 =$ _____. Then model solving the problem by dividing 10 books into 3 stacks. Hold up the book that is left over and say, *This is a remainder. The remainder is one.* Have students repeat chorally. Repeat the procedure using other division problems with small numbers and a remainder (for example, $9 \div 2$, $11 \div 4$, $8 \div 3$). At the end of each problem, hold up however many books are left over, and have students identify the remainder using this sentence frame: **That is the remainder. The remainder is _____.** | **Act It Out** Divide students into pairs, and give each pair 30 counters. On the board, write random division problems in which the dividend is 30 or less. Then have partners model and solve each problem using their counters. After solving each problem, have students identify the remainder (if there is one), using the following sentence frame: **The remainder is _____.** Discuss how interpreting the remainder in a word problem will help them fully understand and apply the problem's solution. | **Synthesis** First, have multilingual groups write their own word problems with solutions that are quotients with remainders. Then, ask students to describe a situation in which a quotient with a remainder should be rounded up to the nearest whole number. Ask groups to write a word problem that reflects the situation. Finally, ask students to describe a situation in which a quotient with a remainder should be rounded down to the nearest whole number. Ask groups to write a word problem that reflects the situation. |

**Teacher Notes:**

NAME _____ DATE _____

# Lesson 6 Multiple Meaning Word
*Interpret Remainders*

Complete the four-square chart to review the multiple meaning word.

| Everyday Use | Math Use in a Sentence |
|---|---|
| Sample answer: The amount that is left over. | Sample sentence: When you divide 7 by 2, you will have a remainder of 1. |
| **Math Use** | **Example From This Lesson** |
| The number that is left after one whole number is divided by another. | Sample answer: Some numbers do not divide evenly. The amount left over is called the remainder. |

**remainder**

Write the correct terms on the lines to complete the sentences.

When you divide 153 by 10, you will have a quotient of _15_ with a remainder of _3_.

153 people were riding vans to a park. Each van holds 10 people each. How many vans would be needed to move all 153 people? _16_

# Lesson 7 Place the First Digit

## English Learner Instructional Strategy

### Collaborative Support: Partners Work/Pairs Check

Pair students and assign Exercises 3–8 in Independent Practice. For the first problem, have one student coach the other in finding the estimate and then the solution. For the second problem, have students switch roles. When pairs have finished the second problem, have them get together with another pair and check answers.

Provide these sentence frames:
**What is your estimate for Exercise \_\_\_\_? Our estimate is \_\_\_\_.**
**How did you round your numbers? We rounded our numbers by \_\_\_\_.**
**What was your solution? Our solution was \_\_\_\_.**

When both pairs have agreed that their solutions are reasonable, ask them to shake hands and continue working in their original pairs for the next two problems. Then have new sets of pairs check answers. Repeat using the established procedure.

### English Language Development Leveled Activities

| Emerging Level | Expanding Level | Bridging Level |
| --- | --- | --- |
| **Academic Vocabulary**<br><br>Explain the meanings of *first*, *second*, and *third*. Label the bases in a picture of a baseball diamond *first*, *second*, and *third*. Then point to each base in order and say, *This is **first** base, this is **second** base, and this is **third** base.* Now, write 367 and label its digits first, second, and third. Point to each digit in order and say, *This is the **first** digit, this is the **second** digit, and this is the **third** digit.* Finally, have students identify the digits of other three-digit numbers as you point to them, using this sentence frame: **The [first/second/third] digit is \_\_\_\_.** | **Number Sense**<br><br>Write these problems on the board: 3)$\overline{28}$, 5)$\overline{452}$, 7)$\overline{904}$, 6)$\overline{631}$, 2)$\overline{14}$. For the first problem, guide students in determining whether the first digit of the quotient should be placed over the ones, tens, or hundreds place of the dividend. Say, *There are 2 tens. That is not enough tens to divide by 3. So, the first digit of the quotient will be in the **ones** place.* Then ask students to determine the first digit of the quotient for the rest of the problems. Have them use this sentence frame: **The first digit goes in the \_\_\_\_ place.** | **Show What You Know**<br><br>Have multilingual groups create a graphic organizer that shows the sequence of steps for solving 618 ÷ 7. Tell students to include the following terms in their graphic organizers: *dividend, divisor, place value, quotient,* and *remainder* Ask a volunteer from each group to present their graphic organizer. |

**Teacher Notes:**

NAME _____ DATE _____

# Lesson 7 Note Taking

## *Place the First Digit*

Read the question. Write words you need help with and research each word. Use your lesson to write your Cornell notes. Write or draw math examples to explain your thinking. Share your examples with a classmate.

| **Building on the Essential Question** | **Notes:** |
|---|---|
| How can you place the first digit? | Sometimes the first digit of the <u>dividend</u> is less than the divisor. You may not be able to <u>place</u> the first digit of the quotient over the first digit of the dividend. |
| | $3\overline{)28}$ |
| | Since 3 is greater than 2, you cannot divide the tens. |
| | Begin division with the <u>ones</u>. Divide the ones place. <u>28</u> ones ÷ 3 = <u>9</u> ones |
| | Place the 9 over the <u>8</u> in the ones place. ⟶ 9 R 1 |
| | Next, multiply: 3 × <u>9</u> ones = <u>27</u> ones    $3\overline{)28}$ |
| | Subtract the 27 from 28.    −27 |
| | <u>28</u> ones − <u>27</u> ones = <u>1</u> one.    1 |
| | Compare to the divisor. |
| | Since 1 is <u>less</u> than 3, 1 is the remainder. |
| | 28 ÷ 3 = <u>9</u> R <u>1</u> |
| **Words I need help with:** | |
| See students' words. | |

**My Math Examples:**

See students' examples.

# Lesson 8 Inquiry/Hands On: Distributive Property and Partial Quotients

## English Learner Instructional Strategy

### Vocabulary Support: Activate Prior Knowledge

Before the lesson, write: *partial product, partial quotient*. Point to *partial product* and have students say the term chorally. Ask, *Is a partial product the whole product or part of the product?* **part** Then point to *partial quotient* and say it. Have students chorally repeat. Ask, *Is a quotient the solution to a multiplication problem or a division problem?* **division** *Do you think a partial quotient is the whole quotient or part of the quotient?* **part**

### English Language Development Leveled Activities

| Emerging Level | Expanding Level | Bridging Level |
|---|---|---|
| **Look, Listen, and Identify** | **Communication Guide** | **Explore Language Structure** |
| Write: *dividend, divisor, quotient, known, unknown.* Discuss how *known* and *unknown* are opposites. Then write this problem and draw an area model that represents it: 24 ÷ 8 = ____. Point to each part of the equation and ask students to correctly identify it as either **dividend, divisor,** or **quotient.** Then have students do the same for the area model. Finally, ask, *Is the dividend known or unknown?* **known** *Is the divisor known or unknown?* **known** *Is the quotient known or unknown?* **unknown** | Remind students that a partial quotient is *part* of a whole quotient. Then pair up students to work on Practice It Exercises 3–6. Have them use this communication guide to help them visualize the area models and discuss the solutions: ____ **(hundreds) divided by** ____ **equals part of the quotient:** ____. ____ **(tens) divided by** ____ **equals part of the quotient:** ____. ____ **(ones) divided by** ____ **equals part of the quotient:** ____. ____ + ____ + ____ = **the whole quotient:** ____. After students have completed their work, have them check answers with another set of partners. | Have students do the Expanding-Level activity, using these sentence frames instead: ____ **goes into** ____ **(hundreds)** ____ **times.** ____ **goes into** ____ **(tens)** ____ **times.** ____ **goes into** ____ **ones** ____ **times. Find the whole quotient by adding the partial quotients:** ____ + ____ + ____ = ____. Point out the idiomatic phrase *"goes into"* and tell students this is a signal phrase for division. Write and say: *24 ÷ 8 = 3; Twenty-four divided by eight equals three;* and *Three goes into twenty-four eight times.* Discuss how these are all ways of saying the same thing. |

**Teacher Notes:**

NAME _____ DATE _____

# Lesson 8 Guided Writing

## Inquiry/Hands On: Distributive Property and Partial Quotients

**How do you divide using the distributive property and partial quotients?**

**Use the exercises below to help you build on answering the Essential Question. Write the correct word or phrase on the lines provided.**

1. Rewrite the question in your own words.
   See students' work.

2. What key words do you see in the question?
   distributive property, partial quotients, divide

3. When you <u>decompose</u> a number, you break the number into different parts.

4. Decompose 648 into different parts according to place value.
   600 + 40 + 8

5. *Partial quotients* is a <u>dividing</u> method in which the <u>dividend</u> is separated into sections that are easy to divide.

6. Complete the equations that will give the partial quotients for the division of 693 ÷ 3.

   $\underline{600} \div 3 = \underline{200}$
   $\underline{90} \div 3 = \underline{30}$
   $\underline{3} \div 3 = \underline{1}$

7. Find the sum of the partial quotients and solve for 693 ÷ 3.

   $\underline{200} + \underline{30} + \underline{1} = \underline{231}$
   $693 \div 3 = \underline{231}$

8. How do you use the distributive property and partial quotients to divide?
   Decompose the dividend using place value. Divide each place value of
   the dividend by the divisor. Then find the sum of the partial quotients.

# Lesson 9 Divide Greater Numbers
## English Learner Instructional Strategy

### Language Structure Support: Tiered Questions

Write the Talk Math problem on the board, setting it up to focus on place values. Then discuss it using the following tiered questions.

**Emerging:** Point to the dividend and ask, *How many digits are in the dividend?* **3** Say, *Is 5 greater than 7?* **No** *Is 5 greater than 9?* **No** *Is 5 greater than 5?* **No** *We can divide each digit by 5. So, how many digits will the quotient have?* **3**

**Bridging:** Say, *Say, The divisor is **less than** the dividend's first digit. So, how many digits will the quotient have?* **3** *Why?* **The quotient will have a digit in the hundreds, tens, and ones places.**

**Expanding:** Ask, *When do the dividend and the quotient have the same number of digits?* **when the divisor is less than each digit of the dividend** *So, how many digits will the quotient of this problem have?* **3**

### English Language Development Leveled Activities

| Emerging Level | Expanding Level | Bridging Level |
|---|---|---|
| **Background Knowledge** | **Developing Oral Language** | **Show What You Know** |
| Draw an arrow pointing down and say, *down.* Then write the problem from Example 1 on the board. Demonstrate solving the problem as shown, and explain each step. Emphasize the term *bring down* each time you perform that step. Then write the problem from Example 2 on the board. Each time you bring down a number, have students say, **bring down** in unison with you. | Write the problem from Example 1 on the board. Demonstrate solving the problem as shown, and explain each step. Emphasize the term *bring down* each time you perform that step. Then use the same procedure to demonstrate solving selected problems from the Reteach worksheet; however, after each "compare" step, ask, *What should I do now?* Have students respond using this sentence frame: **Bring down the ____.** | Have each student write a four-digit by one-digit division problem and then exchange papers with a partner. Tell students to solve the division problem they received. Then ask them to present their problems and solutions orally. Remind them to use the terms *divide, multiply, subtract,* and *bring down* to describe the steps of the process. |

### Multicultural Teacher Tip

In some cultures, mental math is strongly emphasized. Latin American students in particular may skip intermediate steps when performing algorithms such as long division. Whereas US students are taught to write the numbers they will be subtracting in the process of long division, Latin American students will make the calculations mentally and write only the results.

NAME _____ DATE _____

# Lesson 9 Concept Web

## *Divide Greater Numbers*

Use the concept web to identify the place value of each digit.

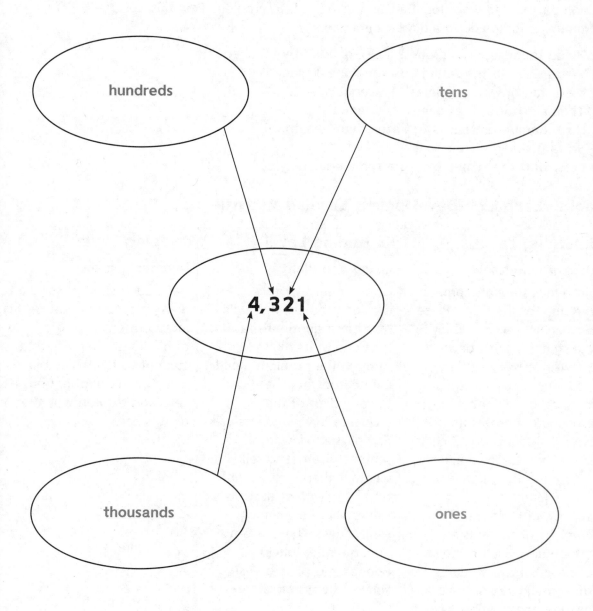

# Lesson 10 Quotients with Zeros
## English Learner Instructional Strategy

### Language Structure Support: Communication Guide

Write *divisor* and its Spanish cognate, *divisor*. Introduce the word, and provide a math example. Utilize other appropriate translation tools for non-Spanish speaking ELs.

Assign partners to work together on the Talk Math problem. Post this communication guide to aid their discussion:

1. Divide [hundreds/tens/ones]. _____ divided by _____ equals _____.
2. Write _____ in the [hundreds/tens/ones] place.
3. Next, multiply. _____ times _____ equals _____.
4. Then subtract. _____ minus _____ equals _____.
5. Now, compare. _____ is [greater than/less than] _____.
6. Bring down the [tens/ones].
7. Last, find the remainder. The remainder is _____.

### English Language Development Leveled Activities

| Emerging Level | Expanding Level | Bridging Level |
|---|---|---|
| **Background Knowledge** | **Listen and Identify** | **Number Sense** |
| Have four students come to the front of the room. Place 5 chairs in a row, and invite the students to sit down. Show thumbs-up and say, *There are **enough** chairs.* Then repeat the above procedure, but this time have 4 students and 3 chairs. Show thumbs-down and say, *There are **not enough** chairs.* Finally, repeat the activity with different numbers of students and other items, such as books or pencils. Have students show thumbs-up if there are enough and thumbs-down if there are not enough. | On the board write a division problem that will result in a quotient that has 0 as one of its digits. Work through the problem aloud. Before applying the divisor to each digit of the dividend, have students use one of the following sentence frames to tell you whether there are *enough* or *not enough* of that place value: There are enough [tens/ones]. **There are not enough [tens/ones]. Repeat the process using similar division problems.** | Have each student solve a division problem that results in a quotient with 0 as one of its digits. Then ask that student to tell which place value of the quotient has a zero and explain why that digit is zero. |

**Teacher Notes:**

NAME _____ DATE _____

# Lesson 10 Vocabulary Chart
## *Quotients with Zeros*

Use the three-column chart to organize the review vocabulary in
this lesson. Write the word in Spanish. Then write the correct terms to
complete each definition.

| English | Spanish | Definition |
|---|---|---|
| **dividend** | dividendo | A number that is being <u>divided</u>. |
| **divisor** | divisor | The number by which the <u>dividend</u> is being divided. |
| **partial quotients** | cocientes parciales | A <u>dividing</u> method in which the <u>dividend</u> is separated into <u>sections</u> that are easy to divide. |
| **quotient** | cociente | The result of a <u>division</u> problem. |
| **remainder** | residuo | The number that is <u>left</u> after one whole number is <u>divided</u> by another. |

# Lesson 11 Solve Multi-Step Word Problems
## English Learner Instructional Strategy

### Vocabulary Support: Build Background Knowledge

Before the lesson, write *equation* and its Spanish cognate, *ecuación*. Introduce the word, and provide a math example. Utilize other translation tools for non-Spanish speaking ELs. Also, explain that *per* means "each" in the Example 1 word problem, and discuss multiple meanings for *change*, which appears in the Example 2 word problem.

Before assigning this lesson's word problems, create the following word webs for signal words/phrases students will encounter. In a *Division* web include *divide equally among, divide equally between, divide into equal groups,* and *divide evenly among.* (Point out these pairs of synonyms among the phrases: *among/ between and equally/evenly*). In a Multiplication web include *times as many*. In an Addition web include *all, everyone, altogether,* and *total.* Discuss the webs, and then post them for students to reference as they work.

### English Language Development Leveled Activities

| Emerging Level | Expanding Level | Bridging Level |
|---|---|---|
| **Making Connections** | **Developing Oral Language** | **Synthesis** |
| Explain multiple meanings for the word *step*. First, slowly take a step forward. As you move, point toward the floor and say, *I walk one step*. Take several more steps and say, *I walk more steps*. Then draw stairs on the board. Point to each horizontal line and say, *This is a step*. Finally, tell students that a step is also one part of a process. Pantomime washing your hands and identify each step as you do so. For example, say, *First step: wet my hands. Next step: rub in soap. Last step: rinse my hands.* | Explain multiple meanings for the word *step*. First, take several steps forward, and say with each step, *I walk one step*. Then draw stairs, and say while pointing to each horizontal line, *This is a step*. Finally, explain that a *step* is also one part of a process. As an example, draw the process for making a smiley face: Step 1, draw a circle; Step 2, draw a circle with one eye; Step 3, draw a circle with two eyes; Step 4, draw a circle with two eyes and a smile. Then point to each circle, and have students identify it as *Step 1, 2, 3* or *4*, as appropriate. | Ask students to think of a favorite sandwich or snack they can make themselves. Tell them to think about the order in which they do the steps for making the snack. Then ask students to write the steps in order. Once they have finished, have them present their instructions to a group of classmates. |

### Teacher Notes:

NAME _____ DATE _____

# Lesson 11 Four-Square Vocabulary
## *Solve Multi-Step Word Problems*

Write the definition for each math word. Write what each word means in your own words. Draw or write examples that show each math word meaning. Then write your own sentences using the words.

| Definition | My Own Words |
|---|---|
| A sentence that contains an equal sign (=), showing that two *expressions* are equal. | See students' examples. |

**equation**

| My Examples | My Sentence |
|---|---|
| Sample answer: $w = 3 \times (10 + 7)$ | Sample sentence: The equation for the cost of three pizzas that are $12 each is $c = 3 \times 12$. |

| Definition | My Own Words |
|---|---|
| The enclosing symbols (), which indicate that the terms within are a unit. | See students' examples. |

**parentheses**

| My Examples | My Sentence |
|---|---|
| Sample answer: $576 = 12 \times (40 + 8)$ | Sample sentence: Using parentheses can help you write an equation for word problems. |

# Chapter 7 Patterns and Sequences

## What's the Math in This Chapter?

### Mathematical Practice 8: Look for and express regularity in repeated reasoning.

Draw an input/output table on the board with 3 rows. Say, *I'm going to write some numbers. I want you to guess what rule I am using.* In the first row, write 2 for the input and 7 for the output. Then in the second row, write 4 for the input and 9 for the output. In the third row, only write the input 6. Give students time to discover that your rule is to add 5. Ask, *What is the last output number?* Have students chorally answer **eleven**.

Ask, *How did you figure out my rule?* Have students turn and talk with a peer. Then discuss as a group. The goal of the discussion is for students to understand that they looked for a number pattern and figured out that a calculation was repeated. Say, *When you guessed my rule, you were looking for patterns and used repeated reasoning.*

Display a chart with Mathematical Practice 8. Restate Mathematical Practice 8 and have students assist in rewriting it as an "I can" statement, for example: **I can find and use patterns to help me solve problems.** Post the new "I can" statement.

## Inquiry of the Essential Question:

### How are patterns used in mathematics?

Inquiry Activity Target: **Students come to a conclusion that equations can be used to represent patterns.**

As an introduction to the chapter, present the Essential Question to students. The inquiry graphic organizer will offer opportunities for students to observe, make inferences, and apply prior knowledge of equations representing the Essential Question. As they investigate, encourage students to draw, write, and collaborate with peers to demonstrate their observations and thinking. Then have students present additional questions they may have to a peer to extend discussions.

Regroup students and restate Mathematical Practice 8 and the Essential Question. Pose questions to reflect on what has been learned to guide students in making connections between the Mathematical Practice and the Essential Question.

NAME _____ DATE _____

# Chapter 7 Patterns and Sequences
## *Inquiry of the Essential Question:*

**How are patterns used in mathematics?**

Read the Essential Question. Describe your observations (I see...), inferences (I think...), and prior knowledge (I know...) of each math example. Write additional questions you have below. Then share your ideas and questions with a classmate.

55, 56, 52, 53, 49, 50, ?
+1 −4 +1 −4 +1

The pattern is add 1, then subtract 4.

I see ...

I think...

I know...

| Input ($a$) | Output ($b$) |
|---|---|
| 4 | 12 |
| 7 | 21 |
| 10 | 30 |
| 13 | ? |

**Rule:** Multiply by 3.
**Equation:** $a \times 3 = b$

I see ...

I think...

I know...

If $x = 6$, what is the value of $y$ in $2 \times (9 + x) = y$?

$$2 \times (9 + x) = y$$
$$2 \times (9 + 6) = y$$
$$2 \times 15 = y$$
$$30 = y$$

I see ...

I think...

I know...

Questions I have...

_____

_____

_____

# Lesson 1 Nonnumeric Patterns

## English Learner Instructional Strategy

### Vocabulary Support: Frontload Academic Vocabulary

Before the lesson, discuss the following lesson terms. First write *numeric* and *nonnumeric*. Explain that *numeric* means "having to do with numbers." Then tell students that *nonnumeric* is an antonym for numeric. Underline the *non-* in *nonnumeric*. Say, *The prefix "non"- means "not." When added to numeric, it forms a word that means "not numeric."* Invite students to share other words that begin with non- (such as, *nonfat, nonstop,* and *nonsense*). Discuss the words' meanings with and without the prefix.

Then write and discuss these multiple-meaning words: *unit, extend, figure,* and *alternates*. Be sure to point out that *alternates*, in the context of this lesson, is a verb with the inflectional ending /s/.

### English Language Development Leveled Activities

| Emerging Level | Expanding Level | Bridging Level |
|---|---|---|
| **Listen and Write** | **Act It Out** | **Show What You Know** |
| Draw a repeating shape pattern, such as square, triangle, square, triangle. Draw a circle around each square-triangle pattern unit, then say, *The pattern repeats. It is the same.* Say, *I am going to extend the pattern.* Add two more complete square-triangle units to the pattern. Invite students to come to the board and extend the pattern further. | Have two girls and two boys stand in a line in the following order: boy, girl, boy, girl. Say, *The pattern unit is: boy, girl.* Point to the first boy and girl in the line. Then say, *The pattern repeats: boy, girl, boy, girl.* Point to each boy and girl as you say the pattern. Then tell students that you want to extend, or add to, the pattern. Ask whether a boy or girl should come next. Have students answer using the following sentence frame: **A ＿＿ comes next.** Invite students to extend the pattern by having additional students join the line. | Have multilingual groups use pattern blocks to create a repeating pattern. Then ask students to identify each group's repeating pattern using the following sentence frame: **The pattern unit is ＿＿.** Now have multilingual groups use the pattern blocks to create a growing pattern. Have a volunteer from each group present and describe their growing pattern to the class. |

**Teacher Notes:**

NAME _____ DATE _____

# Lesson 1 Vocabulary Definition Map
## *Nonnumeric Patterns*

Use the definition map to write a description and list characteristics about the vocabulary word or phrase. Write or draw math examples. Share your examples with a classmate.

**My Math Vocabulary:**

**nonnumeric pattern**

**Characteristics from Lesson:**

A pattern is a sequence of numbers, figures, or symbols that follows a ___rule___ or design.

**Description from Glossary:**

Patterns that do not use numbers.

"Non" means "not", so nonnumeric means __not__ __numeric__.

The nonnumeric pattern shown below is __slice__, __slice__, __whole__.

**My Math Examples:**
See students' examples.

# Lesson 2 Numeric Patterns

## English Learner Instructional Strategy

### Language Structure Support: Sentence Frames

Before the lesson, write and discuss these multiple-meaning words from Math in My World Examples 1 and 2: *rule, even, odd*. Point out that, within the context of this lesson, even and odd are opposites.

Have students complete Exercises 12–15 independently. Then have them check answers with a partner. Give them these sentence frames to use as they discuss: **The rule is _____. Therefore, the unknown is _____.**

Finally, give students these sentences frames to use as they complete Hot Problems Exercise 18:
**The first operation is _____.**
**The second operation is _____.**
**The rule for the pattern is _____.**

### English Language Development Leveled Activities

| Emerging Level | Expanding Level | Bridging Level |
|---|---|---|
| **Academic Vocabulary** | **Listen and Identify** | **Show What You Know** |
| Write the following number pattern on the board: 1, 4, 7, 10. Above and between each pair of numbers, write: +3. Say, *The **rule** is "add 3."* Have students repeat chorally. Next write this number pattern on the board: 20, 15, 10, 5. Above and between each pair of numbers, write: −5. Say, *The **rule** is "subtract 5."* Have students repeat chorally. Finally, have students work in pairs to create their own pattern rule. Ask them to write four numbers following their rule. Then have students share their pattern examples and rules in small groups. | Point to a sign that shows a classroom rule, and say, *A rule tells us what to do.* Then write this number pattern: 2, 6, 10, 14,_____. Say, *A rule tells us how to extend this pattern. What is the rule?* Guide students in identifying that the rule is "add 4." Then ask, *What is the next number in the pattern?* Guide students in identifying that the next number is 18. Repeat the exercise with other number patterns. Have students identify and extend the number patterns using these sentence frames: **The rule is _____. The next number is _____.** | Distribute two index cards to each student. Then ask each student to create a number pattern on one card and write the rule for the pattern on the other card. After students have finished their cards, assign them to multilingual groups. Ask one student in each group to collect, shuffle, and lay out all of the cards from his or her group. Then have students take turns finding matches. |

### Teacher Notes:

NAME _____ DATE _____

# Lesson 2 Concept Web
## *Numeric Patterns*

Use the given rule to write the next number in each numeric pattern shown in the concept web.

2, 8, 14, __20__

1, 7, 13, __19__

3, 9, 15, __21__

**The rule is +6.**

4, 10, 16, __22__

6, 12, 18, __24__

5, 11, 17, __23__

**60** **Grade 4 · Chapter 7** *Patterns and Sequences*

# Lesson 3 Sequences

## English Learner Instructional Strategy

### Collaborative Support: Numbered Heads Together

Write *term* and its Spanish cognate, *término*. Introduce the word, and provide a visual example to support understanding. Utilize other appropriate translation tools for non-Spanish speaking ELs. Then discuss multiple meanings for *term*.

For Independent Practice Exercise 11, divide students into groups of four, and give each student a number from 1 to 4. Have students gather to discuss Exercise 11, agree on the solution, and give an answer. Then read Example 11 aloud, and call out a number (1 to 4) randomly. Have students assigned to that number raise their hands and, when called on, answer for their teams.

### English Language Development Leveled Activities

| Emerging Level | Expanding Level | Bridging Level |
|---|---|---|
| **Developing Oral Language** | **Building Oral Language** | **Synthesis** |
| Write the following number pattern on the board: 50, 100, 150, 200. Point to one of the numbers and say, *This is a term.* Have students repeat chorally. Then circle all four numbers and say, *This is a sequence.* Have students repeat chorally. Continue by writing other number patterns on the board. As you point to one number, have students say, **term**. As you circle the entire series of numbers, have students say, **sequence**. | Write the following sequence on the board: 6, 12, 18, 24, 30. Point to 6 and say, *This is the first term. The first term is 6.* Point to the remaining terms, in order, and have students use the following sentence frames to identify each one: **This is the [second/third/fourth/fifth] term. The [second/third/fourth/fifth] term is _____.** After students have identified the five terms listed, discuss the pattern rule. Then invite volunteers to name the next few terms in the number sequence. | Ask each student to create a sequence of 4 terms based on a pattern. Then ask students to exchange papers with a partner. Have students identify the pattern, name the next four terms in the sequence, and make an observation about the pattern. |

### Multicultural Teacher Tip

As students work together in a variety of collaborative situations, you may notice some ELs seem reluctant to participate. This behavior could be the result of shyness or insecurity due to language issues, but the student may also come from a classroom environment in his or her native culture that did not emphasize group work. The student may be unsure of how to participate or what role to play, worrying that collaboration may be akin to cheating. Be patient and encouraging, and allow the student time to get comfortable with working in a group dynamic.

NAME _____ DATE _____

# Lesson 3 Note Taking
## *Sequences*

Read the question. Write words you need help with and research each word. Use your lesson to write your Cornell notes. Write or draw math examples to explain your thinking. Share your examples with a classmate.

| Building on the Essential Question | Notes: |
|---|---|
| How are sequences used in mathematics? | A term is each number in a <u>numeric</u> <u>pattern</u>. |
| | A sequence is an ordered arrangement of <u>terms</u> that make up a <u>pattern</u>. |
| | A rule is a statement that describes a <u>relationship</u> between numbers or objects. |
| | The number 8 is the first <u>term</u> in the sequence 8, 15, 22, 29, 36. |
| | I can observe the <u>sequence</u> to find the rule. |
| | The rule is add <u>7</u>. |
| **Words I need help with:** | I know to use the rule on the last term of the sequence to <u>extend</u> the pattern. $36 + 7 = $ <u>43</u> |
| See students' words. | I can extend the pattern to find the next three terms: 8, 15, 22, 29, 36, <u>43</u>, <u>50</u>, <u>57</u>. |

**My Math Examples:**

See students' examples.

# Lesson 4 Problem-Solving Investigation
## Strategy: Look for a Pattern

### English Learner Instructional Strategy

### Graphic Support: Graphic Organizer

Before beginning the Learn the Strategy lesson, explain the word *walk-a-thon*. Compare it to *marathon*, a footrace of 26 miles 385 yards.

Show students how to make a table to help them visualize the facts known and look for a pattern in the word problems. Have students model their tables after the one used in Exercise 2. Encourage students to copy the table into their math journals and use it as a model to help them throughout the lesson.

### English Language Development Leveled Activities

| Emerging Level | Expanding Level | Bridging Level |
|---|---|---|
| **Word Knowledge**<br>Find a color pattern in the room, such as the red and white stripes on the American flag. Say, *The stripes make a **pattern**. The **pattern** is red, white, red, white, red, white.* Guide students in identifying other color or shape patterns in the room or on their clothing. Have student describe the pattern using the following sentence frame: **The pattern is ____.** | **Number Sense**<br>Draw the following table:<br><br>Say, *This table shows a pattern. What is the pattern?* Guide students as needed in identifying that the pattern is "add 3." Once students have identified the pattern, ask, *How much money is saved after 4 weeks?* **$12** Invite volunteers to extend the pattern, determining the amount of money saved after 5 weeks, 6 weeks, and so on. | **Public Speaking Norms**<br>Have students work in multilingual groups to write a word problem that can be solved by looking for a pattern. Refer them to exercises in the textbook for examples. Then ask groups to present their word problem and demonstrate its solution. |

| Weeks | 1 | 2 | 3 | 4 |
|---|---|---|---|---|
| Money Saved | $3 | $6 | $9 | |

### Teacher Notes:

NAME _____ DATE _____

# Lesson 4 Problem-Solving Investigation
## STRATEGY: Look for a Pattern

Look for a pattern to solve each problem.

airplane

1. A store sold **48** model airplanes in **August**,
   **58** model airplanes in **September**,
   and **68** model airplanes in **October**.
   Suppose this **pattern continues**.
   **How many** model airplanes will be sold in **December**?

| **Understand**<br>I know:<br><br><br><br>I need to find: | **Solve**<br>Sequence:<br><br>Rule:<br><br>Next term in pattern: |
|---|---|
| **Plan**<br><br>I will look for a ___pattern___ to solve the problem. | **Check** |

2. The **table** shows how many tickets were
   **sold** for the school play **each day**.
   Based on the **pattern**, how many tickets
   will be **sold** on **Friday**?

| Day | Number of Tickets |
|---|---|
| Monday | 312 |
| Tuesday | 316 |
| Wednesday | 320 |
| Thursday | 324 |

| **Understand**<br>I know:<br><br><br><br>I need to find: | **Solve**<br>Sequence:<br><br>Rule:<br><br>Next term in pattern: |
|---|---|
| **Plan**<br><br>I will look for a ___pattern___ to solve the problem. | **Check** |

# Lesson 5 Addition and Subtraction Rules

## English Learner Instructional Strategy

### Sensory Support: Mnemonic Device

Before the lesson, discuss the terms *input* and *output*. Underline the prefixes *in-* in *input* and *out-* in *output*. Say, *The prefix "in"- means "in," and the prefix "out"- means "out."* Discuss why these prefixes make the words input and output opposites.

Then show students the input/output table in Math in My World Example 1. Have students recall what a variable is. Then explain how the table shows different number values for the input variable *x* and the output variable *y*. Briefly discuss how the input/output table "changes" numbers by applying a rule to the input number. Encourage students to visualize the table as a machine: the *input* number is *put into* the machine, combined with other ingredients (the rule), and then *put out of* the machine in a new form.

### English Language Development Leveled Activities

| Emerging Level | Expanding Level | Bridging Level |
|---|---|---|
| **Listen and Identify** Write the following equation on the board: $a + 7 = b$. Point to the letter *a*, and say, *This is a* **variable**. Have students repeat chorally. Repeat the process to identify the letter *b* as a variable. Then circle the entire equation and say, *This is an* **equation**. Have students repeat chorally. Finally, use the procedure outlined above to examine this equation with students: $x - 3 = y$. | **Listen and Identify** Draw this table, and discuss the number pattern it demonstrates: | **Public Speaking Norms** Ask each student to create a rule and write an equation for it. Then have students exchange papers with a partner. Ask students to create an input/output table for the equation they were given. Provide support when needed. Then have students present their completed tables in multilingual groups. Tell them to use the terms: *rule, equation, variable, input,* and *output* in their presentations. |

Expanding Level table:

| Input (*a*) | Output (*b*) |
|---|---|
| 4 | 9 |
| 5 | 10 |
| 6 | 11 |

Point to *a* and say, *This is the* **input variable**. Next point to *b* and say, *This is the* **output variable**. Then ask, What is the rule? **add 5** Write: $a + 5 = b$. Say, *This equation describes the number pattern.*

### Teacher Notes:

NAME _____ DATE _____

# Lesson 5 Note Taking

## Addition and Subtraction Rules

Read the question. Write words you need help with and research each word. Use your lesson to write your Cornell notes. Write or draw math examples to explain your thinking. Share your examples with a classmate.

| Building on the Essential Question | Notes: |
|---|---|
| How are addition and subtraction rules like patterns? | The rule in the pattern below is __subtract 8__. 150, 142, 134, 126, 118 |

**Notes:**

The rule in the pattern below is __subtract 8__.
150, 142, 134, 126, 118

To find the next term, __subtract 8__ from the last term. $118 - \underline{8} = \underline{110}$

An *input* is a quantity that is changed to produce an __output__.

An *output* is the result of an input quantity being __changed__.

| Input (x) | Output (y) |
|---|---|
| 118 | 110 |
| 110 | 102 |

**Words I need help with:**

See students' words.

The first input (x) is __118__.
The first output (y) is __110__.
I know I must subtract 8 from 118 to get __110__.

The second input (x) is __110__.
If I subtract 8 from the input (x) 110, I get an output of __102__.

The numbers change in the same way each time.
The rule is $x - \underline{8} = y$.

**My Math Examples:**

See students' examples.

# Lesson 6 Multiplication and Division Rules
## English Learner Instructional Strategy

### Collaborative Support: Turn and Talk

During Talk Math, have students think about their answers to the question independently. Then allow them to turn and talk to a neighbor about their ideas. Give them these sentence frames to use during their discussion: **An equation and a rule are alike because ____. They are different because an equation has ____ and a rule has ____.** Then ask volunteers to share their answers with the class.

Allow students also to compare answers to Independent Practice Exercises 2–7. Post this communication guide for them to use: **The equation is: ____ [multiplied by/divided by] ____ equals ____. Therefore, the next three output numbers are ____, ____, and ____.**

### English Language Development Leveled Activities

| Emerging Level | Expanding Level | Bridging Level |
|---|---|---|
| **Background Knowledge**<br><br>Write: $c \times 4 = d$. Point to the equation and say, *This is a **multiplication** rule.* Have students repeat chorally. Then draw an input/output table that demonstrates the rule. For Input ($c$), write 2, 3, 4, and for Output ($d$), write 8, 12, 16. Discuss how the equation describes the multiplication pattern. Then write: $r \div 3 = s$. Point to the equation and say, *This is a **division** rule.* Have students repeat chorally. Then draw an input/output table for the rule. For Input ($r$), write 18, 15, 12, and for Output ($s$), write 6, 5, 4. Discuss how the equation describes the division pattern. | **Listen and Identify**<br><br>Write: $a \times 7 = b$. Say, *This is a **multiplication** rule.* Have students repeat chorally. Then draw an input/output table that demonstrates the rule. For Input ($a$), write 7, 8, 9. Then have students help with identifying the output numbers, using this sentence frame: **When the input is ____, the output is ____.** Now write: $c \div 5 = d$. Say, *This is a **division** rule.* Draw an input/output table that demonstrates the rule. For Input ($c$), write 30, 25, 20. Have students help with identifying the output numbers using the same sentence frame they used with the first table. | **Number Game**<br><br>Divide students into pairs, and give each pair a number cube. Post then read aloud these game rules for students to follow: *1. Partners take turns rolling the number cube. 2. Before each roll, partners choose an operation, either multiplication or division. 3. The number cube is rolled to find a number to use with the chosen operation. For example, if division is the operation and a 6 is rolled, the rule is "Divide by 6." 4. Partners race to see who is first to correctly write three equations using the rule.* |

### Teacher Notes:

NAME _____ DATE _____

# Lesson 6 Four-Square Vocabulary

## *Multiplication and Division Rules*

Write the definition for each review math word. Write what each word means in your own words. Draw or write examples that show each math word meaning. Then write your own sentences using the words.

| **Definition** | **My Own Words** |
|---|---|
| An operation on two numbers in which the first number is split into the same number of equal groups as the second number. | See students' examples. |

division

| **My Examples** | **My Sentence** |
|---|---|
| Sample answer: $30 \div 5 = 6$ | Sample sentence: Use division to separate 30 items into 5 equal groups of 6. |

| **Definition** | **My Own Words** |
|---|---|
| An operation on two numbers to find their product. It can be thought of as repeated addition. | See students' examples. |

multiplication

| **My Examples** | **My Sentence** |
|---|---|
| Sample answer: $5 \times 6 = 30$ | Sample sentence: Use multiplication to find the number of items in 5 groups of 6 is 30. |

# Lesson 7 Order of Operations

## English Learner Instructional Strategy

### Vocabulary Support: Multiple Meaning Word

Before the lesson, discuss multiple meanings for the word expression. Say, *A person's face can show how he or she feels.* Model changing your eyes and mouth to show happiness, sadness, and surprise, and then explain that these changes make different facial expressions. Next, discuss how words and actions can be expressions, or ways of showing how one feels. Then say, *expression is also a term you will hear in your Language Arts class.* As an example, discuss idiomatic expressions, such as *Watch out!* and *How cool!* Explain that these are phrases that mean something slightly different from what their words suggest. Finally, discuss the math meaning of *expression*. Encourage students to add this term to their math journals, along with pictures, examples, and notes about its different meanings.

### English Language Development Leveled Activities

| Emerging Level | Expanding Level | Bridging Level |
|---|---|---|
| **Word Knowledge** | **Act It Out** | **Share What You Know** |
| Model a simple restaurant scene. Have two students sit at a table and hold a piece of paper that represents a menu. Walk up to the students with a small pad and say, *May I take your order?* Explain that at a restaurant, an order is what you will eat and drink. Then show students another meaning for order. Write these number series: 4, 1, 2, 5, 3 and 1, 2, 3, 4, 5. Shake your head and draw an X through the first set of numbers. Say, *These are not in order.* Nod your head and circle the second set of numbers. Say, *These are in order.* | Explain the order of operations as follows: First, do operations in parentheses; next, multiply/divide from left to right; and last, add/subtract from left to right. Then make and distribute these three signs: a set of parentheses; a multiplication sign and a division sign with a right-pointing arrow underneath; a plus sign and a minus sign with a right-pointing arrow underneath. Give each sign to a volunteer. Have volunteers arrange themselves at the front of the classroom to correctly show the order of operations. | Pair bridging students with emerging students to review terms from this lesson. Have pairs work together to create a flash card for each of the four operations, that has its name on one side and its symbol on the other. In addition, have them create a flash card with *parentheses* on one side and its symbols on the other. After reviewing the terms and symbols, have students arrange the symbols to show the order of operations. |

## Teacher Notes:

NAME _____ DATE _____

# Lesson 7 Concept Web
## *Order of Operations*

Use the concept web to identify the order to perform the operations in the equation.

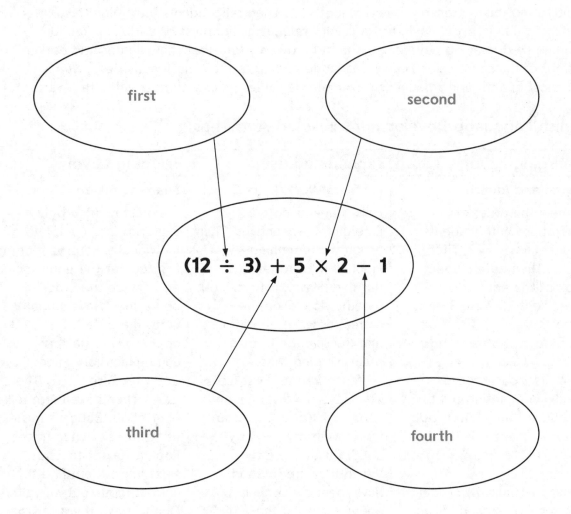

**Word Bank**

first         second         third         fourth

first

second

$(12 \div 3) + 5 \times 2 - 1$

third

fourth

# Lesson 8 Inquiry/Hands On: Equations with Two Operations

## English Learner Instructional Strategy

### Vocabulary Support: Multiple-Meaning Word

Before the lesson, explain that value is a term with multiple meanings. Discuss meanings of value that students encounter in their school subjects. For example, in Reading and Language Arts, students write and talk about personal values, belonging both to themselves and characters in stories. Explain that a personal value is an idea that is very important to someone. Examples include: *honesty – when someone always tells the truth*, *loyalty – when someone is always your friend, even during bad times*, and *love – when someone cares about you very much*. Invite students to tell about what they value in a friend, using this sentence frame: **I value _____.** Then tell students they might also hear the term value in art class. In art, a color's *value* is its lightness or darkness. Reinforce meaning by displaying paint chips from a hardware store that show lighter and darker versions of the same color. Finally, tell students that in their math activities, *value* is a numerical quantity or amount. For example, variables have a number value that varies.

### English Language Development Leveled Activities

| Emerging Level | Expanding Level | Bridging Level |
|---|---|---|
| **Listen and Identify** | **Partners Work/Pairs Check** | **Share What You Know** |
| Review the order of operations with students. Write: **First, _____. Then, _____.** Then have students look at the equation in Practice It Exercise 3. Ask, *What do we do first? Do we add or multiply?* **add** *What symbols show you to add first?* **parentheses** Model completing the first sentence frame: **First, add.** Say, *T equals 4. What is 4 plus 8?* **12** *Now, what should we do? Should we add or multiply?* **multiply** Model the second sentence frame: **Then, multiply.** *What is 12 times 2?* **24** *Good! The variable is s, and s equals 24.* Repeat with similar problems. | Have students do the Emerging-Level activity. Then assign the remaining Practice It Exercises to partners. Have partners work together to decide the correct order of operations and the solution to their problem using these sentence frames: **First, _____. Then, _____. Add _____ and _____. Subtract _____ from _____. Multiply _____ by _____. Divide _____ by _____.** When they are finished, have them check their answers against those of another pair of students. Have the four students work together until they agree on all solutions. | Assign students Apply It Exercise 12 to complete individually. Remind them to first look at the equation and decide the order of operations. Have students write the order of operations using this communication guide: **First, _____ because _____. Then, _____.** Then have them tell a peer about Robert's mistake using this sentence frame: **Robert forgot to _____.** Next, have them find the correct input/output values. Finally, have them rewrite the equation, as directed. Once they are finished, have them teach the solution to an Emerging or Expanding Level peer. |

**Teacher Notes:**

NAME _____ DATE _____

# Lesson 8 Guided Writing

## Inquiry/Hands On: Equations with Two Operations

**How do you solve equations with two operations?**

Use the exercises below to help you build on answering the Essential Question. Write the correct word or phrase on the lines provided.

1. Rewrite the question in your own words.
   See students' work.

2. What key words do you see in the question?
   solve, equations, operations

3. The order of operations are ___rules___ that tell what ___order___ to follow when evaluating an expression.

4. When evaluating an expression, do the ___operations___ in the ___parentheses___ first. ___Multiply___ and ___divide___ in order from left to right, second. ___Add___ and ___subtract___ in order from left to right, last.

5. In the equation $2 \times (n + 7)$, which operation will you perform first?
   addition

6. Rewrite the expression $2 \times (n + 7)$, using $n = 3$.
   $2 \times (3 + 7)$

7. Solve for $m$, when $n = 3$.
   $2 \times (n + 7) = m$.
   $m = $ ___20___

8. How do you solve equations with two operations?
   Follow the order of operations; perform operations in the parentheses first.

# Lesson 9 Equations with Multiple Operations

## English Learner Instructional Strategy

### Language Structure Support: Echo Reading

For Problem Solving Exercises 10–12, pair multilingual students. Have students echo read the problems, taking special care to enunciate each word correctly.

Review English terms from the Exercises which may be unfamiliar such as: *park, fair, tickets, walk, gym class, mile, find, correct, mistake.* Display pictures, photos, or realia to clarify word meaning. Encourage pairs to refer to these visual supports as needed.

### English Language Development Leveled Activities

| Emerging Level | Expanding Level | Bridging Level |
|---|---|---|
| **Developing Oral Language** | **Show What You Know** | **Synthesis** |
| Write this equation: $3 \times (4 + x) - 10 = y$. Point to each operation symbol one at a time. As you point to each symbol, have students name the operation it stands for. **multiplication, addition, subtraction** Next, review the order of operations. Then have students find the value of $y$ when $x = 2$. $y = 8$ Discuss with students how they arrived at their answer. Then repeat the procedure using other equations with multiple operations. | Review the order of operations (first, do operations inside parentheses; next, multiply/divide from left to right; last, add/subtract from left to right). Then write an equation with multiple operations, such as $(13 - x) + 5 \times 2 = y$. Have a volunteer come to the board to write a 1 over the first operation to perform and then name the operation. **subtraction** Then ask additional volunteers to write the number of each remaining step over its operation and name the operation. Repeat the procedure using other equations. | Have multilingual groups work cooperatively to write a word problem that requires multiple operations. Refer them to the lesson for examples, as needed. Then have each group present its problem and challenge the other groups to solve it. |

## Teacher Notes:

NAME _____ DATE _____

# Lesson 9 Multiple Meaning Word
## *Equations with Multiple Operations*

**Complete the four-square chart to review the multiple meaning word or phrase.**

| | |
|---|---|
| **Everyday Use**<br><br>Sample answer: Surgery performed on a patient. | **Math Use in a Sentence**<br><br>Sample sentence: When you combine two groups you use the operation of addition. |
| **Math Use**<br><br>An operation is a mathematical process such as addition (+), subtraction (−), multiplication (×), or division (÷). | **Example From This Lesson**<br><br>Sample answer: In the equation 11 × 3 − 5 two operations (multiplication and subtraction) are used. |

**operation**

**Write the correct symbol, or term on each line to complete the sentence.**

A family bought three pizzas that cost $11 each. They had a coupon for $5 off any order. The equation that gives the total cost of the order uses two operations. The equation is $11 _× 3 _− $5 = $28. The two operations are __multiplication__ and __subtraction__ .

# Chapter 8 Fractions

## What's the Math in This Chapter?

**Mathematical Practice 3: Construct viable arguments and critique the reasoning of others.**

Display 2 circle fraction models. One showing $\frac{1}{2}$ and the other showing $\frac{2}{8}$. Say, *These fractions are the same amount. They are equivalent. Do you agree?* Allow time for students to think. Then have them turn and talk with a peer. Regroup and discuss their observations. The discussion goal should be for students to "construct viable arguments" proving that your fractions are not equivalent.

Ask, *How can I change my fraction models to show equivalent fractions?* change $\frac{2}{8}$ to $\frac{4}{8}$

Discuss with students how they critiqued your reasoning (the fraction models) and provided viable arguments as to why the fractions were not equivalent. Say, *All of you were applying Mathematical Practice 3.*

Display a chart with Mathematical Practice 3. Restate Mathematical Practice 3 and have students assist in rewriting it as an "I can" statement, for example: **I can respond to the problem solving techniques of others.** Post the new "I can" statement.

## Inquiry of the Essential Question:

### How can different fractions name the same amount?

Inquiry Activity Target: **Students come to a conclusion that they can use different models for fractions.**

As an introduction to the chapter, present the Essential Question to students. The inquiry graphic organizer will offer opportunities for students to observe, make inferences, and apply prior knowledge of fractions representing the Essential Question. As they investigate, encourage students to draw, write, and collaborate with peers to demonstrate their observations and thinking. Then have students present additional questions they may have to a peer to extend discussions.

Regroup students and restate Mathematical Practice 3 and the Essential Question. Pose questions to reflect on what has been learned to guide students in making connections between the Mathematical Practice and the Essential Question.

NAME _____  DATE _____

# Chapter 8 Fractions

## *Inquiry of the Essential Question:*

**How can different fractions name the same amount?**

Read the Essential Question. Describe your observations (I see...), inferences (I think...), and prior knowledge (I know...) of each math example. Write additional questions you have below. Then share your ideas and questions with a classmate.

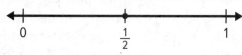

I see ...

I think...

I know...

$$\frac{3}{8} = \frac{3 \times 2}{8 \times 2} = \frac{6}{16}$$

$$\frac{3}{8} = \frac{3 \times 3}{8 \times 3} = \frac{9}{24}$$

I see ...

I think...

I know...

I see ...

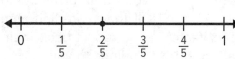

I think...

I know...

$\frac{2}{5} < \frac{1}{2}$ and $\frac{5}{8} > \frac{1}{2}$. So, $\frac{2}{5} < \frac{5}{8}$.

Questions I have...

_____

_____

_____

# Lesson 1 Factors and Multiples

*English Learner Instructional Strategy*

## Language Structure Support: Sentence Frames

Before the lesson, write *pair* and its Spanish cognate, *par*. Introduce the words, and model with realia to support understanding. Utilize other appropriate translation tools for non-Spanish speaking ELs. Then tell students that *pair* has two homophones: *pear* and *pare*. Define the homophones, and point out their spelling differences. Finally, briefly explain what a *factor pair* is, providing math examples.

Provide this sentence frame for the Talk Math question: **I would rather [divide/list the multiples] because \_\_\_\_.** Also, provide this sentence frame for students to explain Exercise 23 to a friend: **Ella [will/will not] say the number 73 because \_\_\_\_.**

## English Language Development Leveled Activities

| Emerging Level | Expanding Level | Bridging Level |
|---|---|---|
| **Word Recognition** | **Building Oral Language** | **Show What You Know** |
| Group students by twos to model the term *pairs*. For each group of two, say, \_\_\_\_ *and* \_\_\_\_ *are a* **pair.** Also, point out things that come in pairs. For example, point to and say, *These are a pair of shoes. These are a pair of socks.* Have students repeat chorally. Then explain that a *factor pair* is the two factors multiplied to find a product. Model listing the factor pairs of 12: 1 and 12, 2 and 6, 3 and 4. Then ask students to find the factor pairs of 25 with a partner. | Explain that a *factor pair* is the two factors multiplied to find a product. Write this on the board four times: \_\_\_\_ × \_\_\_\_ = 24. Tell students there are four factor pairs whose product is 24. Ask volunteers to fill in the blanks to make each equation true. Be sure students use different factor pairs in each equation. **1 and 24; 2 and 12; 3 and 8; 4 and 6** Once the equations are complete, have students identify each factor pair of 24 using the following sentence frame: \_\_\_\_ **and** \_\_\_\_ **are a factor pair of 24.** | Divide students into pairs, and give each pair two number cubes. Ask each student to roll a number. Then have partners use their numbers to make a two-digit number. Finally, have partners work together to find all of the factor pairs of their two-digit number and report back their findings to you. |

## Teacher Notes:

NAME _____ DATE _____

# Lesson 1 Vocabulary Definition Map

*Factors and Multiples*

Use the definition map to write a description and list characteristics about the vocabulary word or phrase. Write or draw math examples. Share your examples with a classmate.

My Math Vocabulary:

## factor pairs

**Characteristics from Lesson:**

A factor is a number that divides a whole number __evenly__ .

**Description from Glossary:**

The two factors that are multiplied to find a product.

A factor is a number that is __multiplied__ by another number.

A number, like 12, can have multiple factor pairs. For example:

$1 \times 12 = 12$
$2 \times \underline{6} = 12$
$3 \times \underline{4} = 12$

My Math Examples:
See students' examples.

# Lesson 2 Prime and Composite Numbers

## English Learner Instructional Strategy

### Vocabulary Support: Read-Aloud/Modeled Talk

Before the lesson, write *prime number* and *composite number* and their Spanish cognates, *número primo* and *número compuesto* on a cognate chart. Introduce the words, using math examples. Utilize other appropriate translation tools for non-Spanish speaking ELs.

During the lesson, read aloud the word problems for Problem Solving Exercises 22–25 to aid students' understanding. Enunciate each word clearly. As needed, define the following words using realia, pantomime, and/or photos: *planting, vegetables, garden, seeds, determine, arrange, rows, quilt, sewing, square, fabric, reversing, result.*

### English Language Development Leveled Activities

| Emerging Level | Expanding Level | Bridging Level |
|---|---|---|
| **Listen and Identify** | **Synthesis** | **Deductive Reasoning** |
| Write these numbers and factor pairs on the board: (7: 1 and 7) (59: 1 and 59) (39: 1 and 39; 3 and 13) (16: 1 and 16, 2 and 8, 4 and 4) Point to the factor pair with the 7. Say, *7 has only one factor pair. So, 7 is a **prime** number.* Repeat for the number 59. Then point to the factor pairs for 39. Say, *39 has more than one factor pair. So, 39 is a **composite** number.* Repeat for the number 16. Finally, randomly point to 7, 59, 39, or 16. Have students correctly identify each number by saying either **prime number/composite number** correspondingly. | Give each student 20 counters. Ask students to divide 15 counters into equal groups, and then discuss that 15 is a composite number. Next have students attempt to divide 7 counters into equal groups, and then discuss that 7 is a prime number. Repeat with various numbers of counters (for example, 6, 11, 13, and 18). Then draw a T-chart, and write the headings *Prime* and *Composite*. Ask volunteers to name and then write numbers that belong in each category under the appropriate heading. | Ask each student to write two riddles—one about a prime number and one about a composite number. Say this riddle as an example: *I am a prime number greater than 25 and less than 30. What prime number am I?* **I am 29** When students have completed their riddles, have them exchange papers with a partner and solve each other's riddles. |

**Teacher Notes:**

NAME _____ DATE _____

# Lesson 2 Vocabulary Cognates

*Prime and Composite Numbers*

Use the Glossary to define the math word in English and in Spanish in the word boxes. Write a sentence using your math word.

| prime number | número primo |
|---|---|
| **Definition**<br>A whole number with exactly two factors, 1 and itself. | **Definición**<br>Número natural que tiene exactamente dos factores: 1 y sí mismo. |
| **My math word sentence:**<br>Sample answer: The number 5 is a prime number. The only factors of 5 are 1 and 5. | |

| composite number | número compuesto |
|---|---|
| **Definition**<br>A whole number that has more than two factors. | **Definición**<br>Número natural con más de dos factores. |
| **My math word sentence:**<br>Sample answer: The number 6 is a composite number. $2 \times 3 = 6$ and $1 \times 6 = 6$. | |

# Lesson 3 Inquiry/Hands On: Model Equivalent Fractions

## English Learner Instructional Strategy

### Vocabulary Support: Background Knowledge

Before the lesson, display the terms *denominator, equivalent fractions,* and *numerator,* along with their Spanish cognates, *denominador, fracciones equivalentes,* and *numerador,* respectively. Explain the meanings of these words and provide math examples. Encourage students to add this information to their math journals.

Then write and say *generate*. Have students chorally repeat. Circle *gen* and underline *ate* in *generate*. Explain that *gen* is a word part meaning "birth" and *-ate* is a word part found at the ends of some verbs; it is a clue that the word *generate* tells an action. Finally, relate the word part meanings to the definition of *generate,* which, within the context of this lesson, is "to produce." Have students recall how they used input/output machines in Chapter 7 "to produce" new numbers according to a pattern. Have them look for patterns with equivalent fractions during the lesson.

### English Language Development Leveled Activities

| Emerging Level | Expanding Level | Bridging Level |
|---|---|---|
| **Non-Transferrable Sounds** | **Number Sense** | **Show What You Know** |
| Write and say: *generate*. Have students chorally repeat. Explain that in English the letter *g* can say /j/ as in *giant* or /g/ as in *guitar*. Explain that when *g* is at the beginning of a word and followed by an *i* or *e*, the *g* often says /j/. Because /j/ is not used in all languages, give students practice saying words with this sound. For example, say *gel*, taking care to pronouncing the /j/ carefully, so that students can hear that it is different from /sh/ or /zh/. Discuss how using /sh/ in place of /j/ can change the meaning of a word, as with *gel* and *shell*. | Review the meanings of *equivalent fractions, numerator,* and *denominator*. Then read aloud the prompt for Talk About It Exercise 1. Have students identify any words in the prompt that are unfamiliar or confusing so that you can clarify them. Then have students respond to the prompt by circling the correct words in this sentence frame: **In equivalent fractions, the [numerator/denominator] divided by the [numerator/denominator] produces the same quotient in both fractions.** Discuss responses as a group. | Have students do the Expanding-Level activity. Then invite partners to work together to generate other true statements about the equivalent fractions shown in the table for Talk About It Exercise 1. (For example: **When the numerator of a fraction is 1, you can find equivalent fractions by multiplying both the numerator and the denominator by the same number.**) Have partners present their statement along with math examples that prove its validity to the class. |

### Multicultural Teacher Tip

Many cultures emphasize the use of decimal numbers over fractions. For this reason, ELs may be unfamiliar with fractions and how they describe the relationship between a part and the whole. You may want to create a chart showing common fractions, their decimal equivalents, and a visual example, such as a shaded circle or rectangle.

NAME _____ DATE _____

# Lesson 3 Vocabulary Chart

*Inquiry/Hands On: Model Equivalent Fractions*

Use the three-column chart to organize the vocabulary in this lesson.
Write the word in Spanish. Then write the correct terms to complete
each definition.

| English | Spanish | Definition |
|---|---|---|
| **denominator** | denominador | The _bottom_ number in a fraction. |
| **equivalent fractions** | fracciones equivalentes | Fractions that represent the _same_ number. |
| **numerator** | numerador | The number _above_ the bar in a fraction. The part of the fraction that tells _how many_ of the equal parts are being _used_. |
| **fraction** | fracción | A number that represents part of a _whole_ or part of a _set_. |

# Lesson 4 Equivalent Fractions

## English Learner Instructional Strategy

### Sensory Support: Pictures/Photographs

Before the lesson, review *numerator, denominator,* and *equivalent,* and their Spanish cognates, *numerador, denominador,* and *equivalente* on a cognate chart. Utilize other appropriate translation tools for non-Spanish speaking ELs. Remind students that *equivalent* is a synonym of *equal.*

Read aloud the word problem for Problem Solving Exercise 16. Discuss the many confusing words in this problem: *deer, does, male* (homophones) *fawns, bucks, spring* (multiple-meaning words). Use pictures and photos to help students understand how the words are used in context.

### English Language Development Leveled Activities

| Emerging Level | Expanding Level | Bridging Level |
|---|---|---|
| **Number Sense** <br> Write $\frac{3}{4}$ on the board. Point to the 3 and say, *Three is the **numerator***. Have students repeat chorally. Then point to the 4 and say, *Four is the **denominator***. Have students repeat chorally. Next, use fraction tiles to model $\frac{3}{4}$ and $\frac{6}{8}$. Show students that the tiles are the same length. Say, *The fractions are **equivalent***. Finally, write: $\frac{3}{4} = \frac{6}{8}$. Point to the equals sign and say, *The **equals sign** shows that the fractions are **equivalent***. Have students say, **equivalent fractions**. | **Building Oral Language** <br> Distribute fraction tiles to students. Then write these equations on the board: $\frac{3}{4} = \frac{\square}{8}$; $\frac{3}{6} = \frac{1}{\square}$. Have students use fraction tiles to find the numerator or denominator that completes each equation. Then ask students to use this sentence frame to name the missing number: **The missing [numerator/ denominator] is ____.** Repeat the activity with different equivalent fractions missing a numerator or denominator. | **Public Speaking Norms** <br> Have students work in pairs. Assign each pair a fraction, and then direct students to find four equivalent fractions. Ask students to draw models to represent all five equivalent fractions on chart paper. Then have each pair explain to the class how they know that the fractions are equivalent. |

**Teacher Notes:**

NAME _____   DATE _____

# Lesson 4 Concept Web

*Equivalent Fractions*

Use the concept web to identify equivalent fractions. Write "true" if the fractions are equivalent and "false" if they are not.

$\frac{2}{3} = \frac{4}{6}$

true

$\frac{4}{5} = \frac{9}{10}$

false

$\frac{1}{3} = \frac{3}{8}$

false

**equivalent fractions**

$\frac{1}{2} = \frac{4}{8}$

true

$\frac{8}{12} = \frac{2}{3}$

true

$\frac{1}{4} = \frac{4}{12}$

false

# Lesson 5 Simplest Form

## English Learner Instructional Strategy

### Collaborative Support: Pairs or Partners

Write *simplify* and *common factor* and their Spanish cognates *simplificar* and *factor común* on a cognate chart. Introduce the words, and provide math examples to aid students' understanding. Utilize other appropriate translation tools for non-Spanish speaking ELs.

During the lesson, assign partners to work together on Problem Solving Exercises 23–26. Discuss the multiple-meaning word *foot*. Tell students that, in this context, *foot* is a measurement equal to 12 inches. Then direct students' attention to the abbreviation, *in.,* in the table, and tell them it means "inches." Finally, read aloud the word problem and chart. Guide students to understand that each number in the chart will be a part of a fraction, specifically the numerator.

### English Language Development Leveled Activities

| Emerging Level | Expanding Level | Bridging Level |
|---|---|---|
| **Word Recognition** Write $\frac{12}{16}$ on the board. Then say, *I will* **simplify**. Write: $\frac{12 \div 4}{16 \div 4} = \frac{3}{4}$. Circle and say, *This is* **simplest form**. Have students repeat chorally. Have students model the equivalent fractions with fraction tiles. Repeat the process using other fractions that can be simplified. | **Listen and Identify** Write on the board: $\frac{9}{15}$. Say, *This fraction is not in its* **simplest form**. Write: $\frac{9 \div 3}{15 \div 3}$ $= \frac{3}{5}$. Circle $\frac{3}{5}$ and say, *The fraction $\frac{3}{5}$ is in* **simplest form**. Have students repeat chorally. Write random fractions on the board and have students identify them correctly by saying either, **The fraction is in simplest form.** Or **The fraction is not in simplest form.** If the fraction is not in simplest form, guide students in simplifying it. | **Public Speaking Norms** Have multilingual groups work together to create a flow chart that shows the steps for simplifying a fraction. Provide each group a fraction to simplify, and ask them to use the fraction as an example in their flow chart. Once groups have completed their flow charts, have them present their charts to the rest of the class, using these terms: *greatest common factor* and *simplest form*. |

### Teacher Notes:

NAME _____   DATE _____

# Lesson 5 Four-Square Vocabulary
## *Simplest Form*

Write the definition for each math word. Write what each word means in your own words. Draw or write examples that show each math word meaning. Then write your own sentences using the words.

| | |
|---|---|
| **Definition**<br>The greatest of the common factors of two or more numbers. | **My Own Words**<br>See students' examples. |
| **My Examples**<br><br>Sample answer: The greatest common factor of 12 and 20 is 4. | **My Sentence**<br><br>Sample sentence: Using the greatest common factor you can find the simplest form of a fraction. |

**greatest common factor**

| | |
|---|---|
| **Definition**<br>A fraction in which the numerator and denominator have no common factor greater than 1. | **My Own Words**<br>See students' examples |
| **My Examples**<br><br>Sample answers: $\frac{2}{3}, \frac{1}{2}, \frac{4}{5}, \frac{3}{4}$ | **My Sentence**<br><br>Sample sentence: The fraction $\frac{3}{5}$ is in simplest form. |

**simplest form**

Grade 4 • **Chapter 8** *Fractions* **73**

# Lesson 6 Compare and Order Fractions
## English Learner Instructional Strategy

### Language Structure Support: Communication Guide

Before the lesson, write *common multiple* and its Spanish cognate, *múltiplo comú* on a cognate chart. Introduce the term, and provide a math example to aid students' understanding. Utilize other appropriate translation tools for non-Spanish speaking ELs. Then discuss multiple meanings for *common*. Ask, *What does **common** mean in this lesson?* **shared**

During the lesson, have partners discuss the Talk Math question, using this communication guide:

**First, find a common multiple of the fractions' _____.**
**The common multiple is _____.**

**Next, make an equivalent fraction for _____.**
**The equivalent fraction is _____.**

**Finally, compare these two fractions: _____ and _____.**
**_____ is the greater fraction.**

### English Language Development Leveled Activities

| Emerging Level | Expanding Level | Bridging Level |
|---|---|---|
| **Word Recognition** | **Number Sense** | **Public Speaking Norms** |
| Write this series of numbers on the board: 1, 5, 7, 12. Point to the 1 and say, *This number, 1, is the least.* Have students repeat chorally. Then point to the 12 and say, *This number, 12, is the greatest.* Have students repeat chorally. Now move your hand beneath the numbers from left to right and say, *The numbers are in order from least to greatest.* Have students repeat chorally. Repeat the exercise with a series of fractions that increase in value, such as $\frac{1}{3}$, $\frac{4}{6}$ and $\frac{3}{3}$. Invite students to point to the greatest and least fraction in the series. | Review how to find the least common multiple to create equivalent fractions. Then have students pair up. Ask each student to roll a number cube. Tell pairs that the number they roll will create a fraction; the highest number will be the denominator, and the lowest number will be the numerator. Then have students repeat the process to create two more fractions. Finally, have students write the three fractions in order from *least to greatest.* Repeat the exercise, having students order the numbers *greatest to least.* | Have multilingual groups work together to write instructions for ordering a series of fractions with the same denominator and for ordering a series of fractions with the same numerator. Encourage students to draw a number line with the fractions plotted to support their instructions. Then ask the groups to present their instructions to the class. |

**Teacher Notes:**

NAME _____ DATE _____

# Lesson 6 Note Taking

## Compare and Order Fractions

Read the question. Write words you need help with and research each word. Use your lesson to write your Cornell notes. Write or draw math examples to explain your thinking. Share your examples with a classmate.

| Building on the Essential Question | Notes: |
|---|---|
| How do I compare and order fractions? | In the fraction $\frac{5}{8}$, the numerator is __5__ and the denominator is __8__.<br><br>To compare the fractions $\frac{2}{3}$ and $\frac{5}{8}$, you must first find equivalent fractions, so that the ___denominators___ are the same.<br><br>Find the least common multiple before finding equivalent fractions.<br><br>The multiples of 3 are: 3, 6, 9, 12, __15__, __18__, __21__, ⓸(24).<br><br>The multiples of 8 are: 8, 16, (24), 32, __40__, __48__, __56__, __64__.<br><br>The least common multiple of 3 and 8 is __24__. |
| **Words I need help with:**<br>See students' words. | Next, find equivalent fractions with a denominator of 24.<br><br>$\frac{2}{3} = \frac{16}{24}$ and $\frac{5}{8} = \frac{15}{24}$<br><br>Compare the fractions. $\frac{16}{24} > \frac{15}{24}$, so $\frac{2}{3} \bigcirc> \frac{5}{8}$. |

**My Math Examples:**

See students' examples.

# Lesson 7 Use Benchmark Fractions to Compare and Order

## English Learner Instructional Strategy

### Collaborative Support: Turn and Talk

Ask a volunteer to read aloud the Talk Math prompt. Then have students think about the prompt independently. Encourage them to make notes about their explanations in their math journals. Then allow them time to share their ideas with a neighbor. After students have had a few minutes to practice their explanations, ask all of them to stand up. Call on a volunteer to share his or her explanation with the class and then sit down, along with any students who have similar responses. Continue until all students are seated. Discuss all responses.

Have students work with a bilingual peer or aide to read aloud, discuss, and solve Problem Solving Exercises 14–17.

### English Language Development Leveled Activities

| Emerging Level | Expanding Level | Bridging Level |
|---|---|---|
| **Number Recognition** Model the fraction $\frac{1}{2}$ by drawing a number line, using fraction tiles, and using fraction circles. Each time you model the fraction, say, *This is* **one half**. Have students repeat chorally. Then on the number line you drew, mark the fractions $\frac{1}{4}$ and $\frac{3}{4}$. Point to $\frac{1}{4}$ and say, *The fraction $\frac{1}{4}$ is **less than** the fraction $\frac{1}{2}$.* Finally, point to $\frac{3}{4}$ and say, *The fraction $\frac{3}{4}$ is **greater than** the fraction $\frac{1}{2}$.* Have students place fraction tiles and circles corresponding on the number line to visualize the benchmark fractions. | **Synthesis** Distribute fraction tiles to the students. Have each student model the fraction $\frac{1}{2}$. Then write the fraction $\frac{2}{3}$. Have students model $\frac{2}{3}$. and tell whether it is less than or greater than $\frac{1}{2}$. **greater than** Invite a volunteer to write this on the board: $\frac{2}{3} > \frac{1}{2}$. Then repeat the exercise, having students compare other fractions to the benchmark fraction $\frac{1}{2}$. | **Cultural Connections** Have each student think of a food item from their culture and then write a real-world word problem about the food item. Tell students the problem should involve comparing three fractions. Direct them to the Problem Solving Exercises for examples. Then have students exchange problems with a partner and solve their partner's problem. Afterward, discuss how they determined which type of food to use as the basis for their word problems. |

**Teacher Notes:**

NAME _____ DATE _____

# Lesson 7 Guided Writing

## *Use Benchmark Fractions to Compare and Order*

**How do you use benchmark fractions to compare and order?**

Use the exercises below to help you build on answering the Essential Question. Write the correct word or phrase on the lines provided.

1. Rewrite the question in your own words.
   See students' work.
   _____

2. What key words do you see in the question?
   benchmark, fractions, compare, order,
   _____

3. A _fraction_ is a number that represents part of a whole or part of a set.

4. Benchmark fractions are _common_ fractions that are used for _estimation_.

5. Compare. $\frac{1}{2} \,\text{\textcircled{>}}\, \frac{3}{8}$

   | $\frac{1}{8}$ | $\frac{1}{8}$ | $\frac{1}{8}$ |
   |---|---|---|

   | $\frac{1}{2}$ |
   |---|

6. Compare. $\frac{1}{2} \,\text{\textcircled{<}}\, \frac{3}{5}$

   | $\frac{1}{2}$ |
   |---|

   | $\frac{1}{5}$ | $\frac{1}{5}$ | $\frac{1}{5}$ |
   |---|---|---|

7. Compare. $\frac{3}{8} \,\text{\textcircled{<}}\, \frac{3}{5}$

8. How do you use benchmark fractions to compare and order?
   Use the benchmark fraction $\frac{1}{2}$. Compare each fraction to the
   benchmark fraction before comparing them to each other.

# Lesson 8 Problem-Solving Investigation
# Strategy: Use Logical Reasoning
## English Learner Instructional Strategy

### Graphic Support: Anchor Chart

Display the following framework for solving the Apply the Strategy
Exercises 1 and 2. Pair students and have them reference the anchor
chart as they work on the exercises.

1. **Understand:** Highlight the facts that are known.

2. **Plan:** Use logical reasoning to solve the problem.

3. **Solve:** Use this least common multiple as a denominator for all the
fractions: _____. Then compare the fractions and order them *least to
greatest*: _____, _____, and _____. Logically order the amounts/distances
*least to greatest*: _____, _____, and _____. Match the fraction to the
logically ordered amounts/distances *least to greatest*.

4. **Check:** The amounts/distances match the clues. My solution is correct.

### English Language Development Leveled Activities

| Emerging Level | Expanding Level | Bridging Level |
|---|---|---|
| **Background Knowledge** | **Developing Oral Language** | **Show What You Know** |
| Help students with My Homework Exercise 1 by defining and then displaying a photo of each of these locations from the word problem: *library, mall, bank.* Then read aloud the word problem. Ask, *Which location had the least art?* **bank** *Which location had the most art?* **library** Invite a volunteer to place the photos of the library, mall, and bank in order *least to greatest*. Then have students find the fraction of art sold at each location. Write each fraction on a sticky note, and have students order the fractions *least to greatest*. Finally, have students place each fraction on the correct photo. | Help students with My Homework Exercise 1 by defining terms that may be unfamiliar, such as: *artwork, displayed, location, library, mall,* and *bank*. Then read aloud the word problem with students. Have students use the clues to tell how much art is at each location. Provide these sentence frames: _____ **has some art.** _____ **has more art.** _____ **has the most art.** Then have students order the fractions *least to greatest* and assign the corresponding fractions to the locations. Direct students to use this sentence frame to report back: **The [mall/library/bank] has _____ of the artwork.** | Divide students into multilingual groups, and assign each group a My Homework Problem Solving exercise to solve using logical reasoning. Once groups have solved their problems, have them create a graphic organizer that illustrates the steps they took to solve it. Then ask each group to present their problem and graphic organizer to the rest of the class. |

**Teacher Notes:**

NAME _____ DATE _____

# Lesson 8 Problem-Solving Investigation
## STRATEGY: Use Logical Reasoning

Use logical reasoning to solve each problem.

1. Sophia is making a salad with **tomatoes**, **cucumbers**, and **mozzarella** **cheese**.
   Use the clues to find the **amount** of **each ingredient**.
   The amounts are $\frac{3}{6}$ cup, $\frac{2}{5}$ cup, and $\frac{3}{4}$ cup.
   There is **less** amount of tomatoes **than** cucumbers.
   There is **less** amount of cheese **than** tomatoes.

| Understand | Solve |
|---|---|
| I know: <br><br> I need to find: | |
| **Plan** | **Check** |
| | |

2. **Mason** walked on **Monday**, **Wednesday**, and **Friday**.
   Use the clues to find **how far** he (Mason) walked **each day**.
   The distances were $\frac{6}{8}$ mile, $\frac{1}{4}$ mile, and $\frac{1}{6}$ mile.
   He did **not** walk the **farthest** on Monday.
   He walked **less** on Friday **than** Monday.

| Understand | Solve |
|---|---|
| I know: <br><br> I need to find: | |
| **Plan** | **Check** |
| | |

# Lesson 9 Mixed Numbers

## English Learner Instructional Strategy

### Language Structure Support: Anchor Chart

Before the lesson, write *mixed number* and its Spanish cognate, *número mixto* on a cognate chart. Introduce the term, and provide a math example to support understanding. Utilize other appropriate translation tools for non-Spanish speaking ELs. Then model saying mixed numbers. Write this on the board: $1\frac{1}{2}$. Say, *This stands for "one and one half."* As you say "one" point to the 1, and as you say "one half" point to the $\frac{1}{2}$. Then write: $2\frac{1}{4}$. Say, *This stands for "two and one fourth."* As you say "two" point to the 2, and as you say "one fourth" point to the $\frac{1}{4}$. Clearly enunciate the ending /th/sound in *fourth*. Repeat the process with other mixed numbers. Then have students help you create a poster that shows a variety of mixed numbers in both numeric and word form.

### English Language Development Leveled Activities

| Emerging Level | Expanding Level | Bridging Level |
|---|---|---|
| **Word Recognition** | **Number Recognition** | **Show What You Know** |
| Write 7 on the board. Point to it and say, *Seven is a whole number.* Have students repeat chorally. Then write $\frac{2}{3}$ on the board. Point to it and say, *Two-thirds is a fraction.* Have students repeat chorally. Finally, write $5\frac{1}{6}$ on the board. Point to it and say, *Five and one-sixth is a mixed number.* Have students repeat chorally. Write other examples of each kind of number. Then ask students to write their own examples of a whole number, a fraction, and a mixed number on write-on/wipe-off boards. Check students' work to evaluate their understanding. | Write a variety of whole numbers, fractions, and mixed numbers on the board. Tell students to raise their hands when you point to a mixed number, as you randomly point to numbers on the board. Then have students use fraction tiles to model mixed numbers. | Have students work in multilingual groups. Assign each group a mixed number. Then ask students to write an equation that represents the mixed number as a sum of whole numbers and fractions. (For example: $3\frac{1}{4} = 1 + 1 + 1 + \frac{1}{4}$) Finally, have groups use fraction tiles or fraction circles to model their mixed number. |

**Teacher Notes:**

NAME _____  DATE _____

# Lesson 9 Vocabulary Definition Map
## *Mixed Numbers*

Use the definition map to write a description and list characteristics
about the vocabulary word or phrase. Write or draw math examples.
Share your examples with a classmate.

**My Math Vocabulary:**

> **mixed number**

**Characteristics from Lesson:**

> _Whole_ _numbers_ are the
> numbers 0, 1, 2, 3, 4, ...

**Description from Glossary:**

> A number that has a whole
> number part and a fraction
> part.

> A _fraction_ is a number
> that represents part of a
> whole or part of a set.

> A mixed number represents
> an amount _greater_ than 1.

**My Math Examples:**

See students'
examples.

# Lesson 10 Mixed Numbers and Improper Fractions

## English Learner Instructional Strategy

### Vocabulary Support: Utilize Resources

Before the lesson, have Spanish-speaking students refer to the Glossary for the Spanish definition of *improper fraction*. If necessary, explain how the glossary is organized. Point out that English math terms are alphabetized and defined on the left side of a page and that their Spanish translations are on the right side. Then write: *proper/improper*. Tell students that *proper* means "correct or appropriate." Then say, *Proper and improper are antonyms because they have opposite meanings.* Underline the *im-* in *improper*. Explain that this prefix means "not" and that, when added to *proper,* it forms a word that means "not proper." Invite students to brainstorm other words with *im-,* such as: *impractical, imperfect,* and *impossible.* Discuss the words' meanings with and without the prefix.

### English Language Development Leveled Activities

| Emerging Level | Expanding Level | Bridging Level |
|---|---|---|
| **Word Recognition** Write $\frac{10}{7}$ on the board. Beside it write: 10 > 7. Say, *10 is greater than 7,* as you point to the numerator and then the denominator. Then write $\frac{10}{7}$ and $1\frac{3}{7}$. Circle the fraction and say, *This is an improper fraction.* Have students repeat chorally. Circle the mixed number and say, *This is a mixed number.* Have students repeat chorally. Repeat this process using other examples of improper fractions. | **Number Recognition** Say, *If a fraction's numerator is greater than its denominator, the fraction is an improper fraction. If a fraction's numerator is less than its denominator, the fraction is a proper fraction.* Write a variety of proper and improper fractions on the board. Have students identify each fraction as proper or improper, using this sentence frame: **The fraction is [proper/improper] because _____.** | **Show What You Know** Place random amounts of fraction tiles in bags that comprise a mixed number. Give one bag to each student. Have the students write the mixed number and improper fraction the tiles represent on a write-on/wipe-off board. Direct students to then pair up with a partner and describe their mixed number and improper fraction. Have the students rotate the fraction tile bags and repeat the activity. |

### Teacher Notes:

NAME _____ DATE _____

# Lesson 10 Note Taking
## *Mixed Numbers and Improper Fractions*

**Read the question. Write words you need help with and research each word. Use your lesson to write your Cornell notes. Write or draw math examples to explain your thinking. Share your examples with a classmate.**

| **Building on the Essential Question** | **Notes:** |
|---|---|
| How do I convert between mixed numbers and improper fractions? | A fraction with a numerator that is greater than or equal to the denominator is an <u>improper</u> <u>fraction</u>. <br><br> A number that has a whole number part and a fraction part is a <u>mixed</u> <u>number</u>. <br><br> An improper fraction can be written as a <u>mixed</u> number. <br><br> A mixed number can be written as an <u>improper</u> fraction. <br><br> These are equivalent fractions for one whole. <br><br> $1 = \dfrac{1}{1} = \dfrac{2}{2} = \dfrac{3}{3} = \dfrac{4}{4} = \dfrac{5}{5} = \dfrac{6}{6}$ <br><br> I can use the equivalent fractions of one whole to write a mixed number as an improper fraction. <br><br> $1\dfrac{1}{3} = 1 + \dfrac{1}{3} = \dfrac{3}{3} + \dfrac{1}{3} = \dfrac{4}{3}$ <br><br> I can use the equivalent fractions of one whole to write an improper fraction as a mixed number. <br><br> $\dfrac{8}{5} = \dfrac{5}{5} + \dfrac{3}{5} = 1 + \dfrac{3}{5} = 1\dfrac{3}{5}$ |
| **Words I need help with:** <br> See students' words. | |

**My Math Examples:**

See students' examples.

# Chapter 9 Operations with Fractions

## What's the Math in This Chapter?

### Mathematical Practice 5: Use appropriate tools strategically.

Use the Virtual Manipulatives to display four $\frac{1}{6}$-fraction tiles with plenty of white space between them. Distribute write-on/wipe-off boards and ask students to write an addition sentence for the tiles shown. $\frac{1}{6} + \frac{1}{6} + \frac{1}{6} + \frac{1}{6} = \frac{4}{6}$ Have students display their addition sentences to you and evaluate their thinking.

Discuss and model the correct addition sentence as well as any incorrect addition sentences as a group. (Some students may have added the denominator.) Then group the first three $\frac{1}{6}$-fraction tiles together in the Virtual Manipulatives and move the fourth tile away. Ask students to write a new addition sentence for the model. $\frac{3}{6} + \frac{1}{6} = \frac{4}{6}$ Have students display their addition sentences and discuss as a group. Say, *We used fraction tiles as tools to model the sums of fractions.*

Display a chart with Mathematical Practice 5. Restate Mathematical Practice 5 and have students assist in rewriting it as an "I can" statement, for example: **I can use tools to better understand fractions.** Post the new "I can" statement.

## Inquiry of the Essential Question:

### How can I use operations to model real-world fractions?

Inquiry Activity Target: **Students come to a conclusion that different models can be used to solve problems about fractions.**

As an introduction to the chapter, present the Essential Question to students. The inquiry graphic organizer will offer opportunities for students to observe, make inferences, and apply prior knowledge of models representing the Essential Question. As they investigate, encourage students to draw, write, and collaborate with peers to demonstrate their observations and thinking. Then have students present additional questions they may have to a peer to extend discussions.

Regroup students and restate Mathematical Practice 5 and the Essential Question. Pose questions to reflect on what has been learned to guide students in making connections between the Mathematical Practice and the Essential Question.

NAME _____ DATE _____

# Chapter 9 Operations with Fractions
## *Inquiry of the Essential Question:*

**How can I use operations to model real-world fractions?**

Read the Essential Question. Describe your observations (I see...), inferences (I think...), and prior knowledge (I know...) of each math example. Write additional questions you have below. Then share your ideas and questions with a classmate.

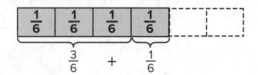

$$\frac{3}{6} \quad + \quad \frac{1}{6}$$

$$\frac{3}{6} + \frac{1}{6} = \frac{4}{6}$$

I see ...

I think...

I know...

$$3\frac{2}{5} = \frac{5}{5} + \frac{5}{5} + \frac{5}{5} + \frac{1}{5} + \frac{1}{5}$$

$$= \frac{5 + 5 + 5 + 1 + 1}{5} = \frac{17}{5}$$

$$1\frac{1}{5} = \frac{5}{5} + \frac{1}{5} = \frac{5 + 1}{5} = \frac{6}{5}$$

$$\frac{17}{5} - \frac{6}{5} = \frac{11}{5} \text{ or } 2\frac{1}{5}$$

I see ...

I think...

I know...

$$\frac{7}{8} = 7 \times \frac{1}{8}$$

I see ...

I think...

I know...

Questions I have...

_____

_____

_____

Grade 4 • **Chapter 9** *Operations with Fractions* **79**

# Lesson 1 Inquiry/Hands On: Use Models to Add Like Fractions

## English Learner Instructional Strategy

### Sensory Support: Pictures

Before the lesson, write *alike* and *different* on the board. Point to each word and pronounce it. Have students chorally repeat. Ask students to define *different* using their own words such as: **not the same.** Then explain that *alike* and *different* are opposites. Use playing cards, such as those for a game of Old Maid or Concentration, to further illustrate. Hold up two cards with the same picture and ask, *Are these pictures different?* **no** Say, *That's right! They are not different; they are the same—they are alike.* Invite students to tell how the pictures are alike, using tiered sentence frames, such as the following:

Emerging: Ask, *What is alike?* **the ____s/es**

Expanding: Ask, *What is alike?* **Both have ____.**

Bridging: Ask, *How are they alike?* **They are alike because both have ____.**

Have students preview the *like fractions* in the lesson. Say, *Like fractions are alike because they have the same denominator.*

## English Language Development Leveled Activities

| Emerging Level | Expanding Level | Bridging Level |
|---|---|---|
| **Choral Responses** Write: $\frac{1}{8}$. Then write *unit fraction* and underline *unit*. Say, *A unit is always one of something.* Write: *unit = 1.* As an example, help students recall that a unit square is one square unit. Then say, *A unit fraction is also one of something.* Circle $\frac{1}{8}$ and say, *This is a unit fraction.* Then point to the numerator and say, *This is one-eighth.* Then write other examples of unit fractions, such as $\frac{1}{4}$, $\frac{1}{3}$, and $\frac{1}{5}$. Have students chorally identify each fraction with you, using this sentence frame: **This is a unit fraction. This is one-____.** | **Look, Listen and Identify** Write $\frac{3}{8}$. Then draw a pie chart that represents the fraction. Ask, *How many parts does the whole pie have?* **8** Point back to the fraction and ask, *How many parts does the whole have?* **8** Next, write: $\frac{1}{8}$, $\frac{5}{8}$, $\frac{7}{8}$. Say, *These are like fractions. How are the fractions alike?* **They have the same denominator.** *Yes, they have the same denominator; they all show parts of the same whole.* Now, read Write About It Exercise 12 with students. Discuss how the denominator does not change when adding like fractions because each fraction represents part of the same whole. | **Signal Words** Write this sentence frame on the board: **You can consider/think of ____ as ____.** Tell students that *consider* and *think of* are synonyms and that each is used to signal a comparison. Model using the sentence frame to compare a fraction with one of its decomposed forms. Say, *You can consider $\frac{3}{8}$ as $\frac{1}{8}$ plus $\frac{2}{8}$. Is $\frac{3}{8}$ the same as $\frac{1}{8}$ plus $\frac{2}{8}$?* **yes** *You can think of $\frac{5}{6}$ as $\frac{2}{6}$ plus $\frac{3}{6}$. Is $\frac{5}{6}$ the same as $\frac{2}{6}$ plus $\frac{3}{6}$?* **yes** Encourage students to use the sentence frames to help them discuss fractions and their decomposed forms. |

## Teacher Notes:

NAME _____ DATE _____

# Lesson 1 Vocabulary Definition Map
*Inquiry/Hands On: Use Models to Add Like Fractions*

Use the definition map to write a description and list characteristics about the vocabulary word or phrase. Write or draw math examples. Share your examples with a classmate.

**My Math Vocabulary:**

## like fractions

**Characteristics from Lesson:**

A denominator is the bottom number in a fraction.

**Description from Glossary:**

Fractions that have the same denominator

The word "like" indicates that the fractions are of the same form. The form is the denominator.

When adding like fractions, the sum has the same denominator as the addends.

**My Math Examples:**

See students' examples.

# Lesson 2 Add Like Fractions

## English Learner Instructional Strategy

### Collaborative Support: Turn and Talk

Before the lesson, write *like fractions* on the board. Circle like and tell students that this is a multiple-meaning word. Say, *If I **like** a person, I think that person is nice. If I **like** a food, I think that food tastes good.* Then say, *In the math term **like fractions**, the word **like** means something different. What does it mean?* Have students turn and talk to a neighbor about its meaning, recalling what they learned about *like fractions* in Lesson 1. **It means "similar."** Point out that *similar* and *like* are synonyms. Then ask, *What makes **like fractions** similar?* **They have the same denominator.**

### English Language Development Leveled Activities

| Emerging Level | Expanding Level | Bridging Level |
|---|---|---|
| **Listen and Identify** Write $\frac{1}{6}$ and $\frac{3}{6}$ on the board. Ask students to point to the denominator in each fraction. Say, *The denominator is 6.* Have students repeat chorally. Then circle the denominators in the fractions and say, *The denominators are the same. So, these fractions are **like fractions**.* Write other pairs of fractions on the board. Have students show thumbs-up if the pairs are like fractions and thumbs-down if the pairs are not like fractions. | **Number Sense** Write $\frac{2}{4} + \frac{1}{4} = \frac{\square}{\square}$ on the board. Circle the numerators and say, *Add the numerators.* Write 3 as the numerator in the solution. Then circle the denominators and say, *Keep the same denominator.* Write 4 as the denominator in the solution. Then model the addition using fraction tiles or fraction circles. Finally, ask volunteers to write addition sentences using like fractions and invite classmates to the board to solve them. | **Show What You Know** Distribute fraction tiles and fraction circles to the students. Have each student create a simple addition problem with like fractions. Then tell students to trade papers with a partner and model each other's addition problem using the fraction tiles or fraction circles. Have students simplify the sum, if necessary. |

### Teacher Notes:

NAME _____ DATE _____

# Lesson 2 Vocabulary Chart
## *Add like Fractions*

Use the three-column chart to organize the review vocabulary in this lesson. Write the word in Spanish. Then write the correct terms to complete each definition.

| English | Spanish | Definition |
|---|---|---|
| **denominator** | denominador | The <u>bottom</u> number in a fraction. In $\frac{5}{6}$, <u>6</u> is the denominator. |
| **numerator** | numerador | The number <u>above</u> the bar in a fraction; the part of the fraction that tells how many of the equal parts are <u>being used</u>. |
| **simplest form** | mínima expresión | A fraction in which the numerator and denominator have <u>no</u> common factor greater than 1. $\frac{3}{5}$ is the simplest form of $\frac{6}{10}$. |
| **greatest common factor** | máximo común divisor | The <u>greatest</u> of the common factors of two or more numbers. The greatest common factor of 12, 18, and 30 is <u>6</u>. |
| **like fractions** | fracciones semejantes | Fractions that have the <u>same</u> denominator. $\frac{1}{5}$ and $\frac{2}{5}$ |

Grade 4 • **Chapter 9** *Operations with Fractions* **81**

# Lesson 3 Inquiry/Hands On: Use Models to Subtract Like Fractions

## English Learner Instructional Strategy

### Vocabulary Support: Activate Prior Knowledge

Before the lesson, have students recall what they know about adding like fractions. Write this sentence frame: **To add like fractions, find the sum of the ____s and keep the ____s the same.** Have students turn and talk to a neighbor about how best to complete the sentence frame. Encourage students to write their responses in their math journals. Then discuss the correct response as a class. **To add like fractions, find the sum of the numerators and keep the denominators the same.**

Then ask, *What is the inverse operation of addition?* **subtraction** Say, *Addition and subtraction are opposites. So, there are opposite rules for adding and subtracting like fractions.* Use the addition rule students wrote to make this sentence frame: **To subtract like fractions, find the ____ of the numerators and keep the denominators the same.** Ask, *What is the opposite of sum?* **difference.**

### English Language Development Leveled Activities

| Emerging Level | Expanding Level | Bridging Level |
| --- | --- | --- |
| **Academic Vocabulary**<br><br>Write and say: *separate*. Have students chorally repeat. Then model how *separate* means "remove" or "move away." For example, show a set of fraction tiles. Say, *I will separate 3 tiles from the group.* Count 3 tiles and then remove them to a different place. Say, *I separated the tiles. I removed the tiles. I moved the tiles away.* Have students demonstrate their understanding using their own sets of tiles. For example, say, *Separate 5 fraction tiles.* Have students respond by counting 5 fraction tiles and removing them to a different place. | **Act It Out**<br><br>Pair students and give each pair a construction paper circle with lines drawn on it, dividing the circle into eighths. Invite pairs to decorate their circle to look like a pizza. Next, have them cut their pizza into eighths, along the circle's dividing lines. Tell them that each eighth of their pizza is called a *slice*. Then read aloud Apply It Exercise 12. Help students identify and define any unfamiliar words. Then have pairs act out the word problem using their pizza. Ask them to complete this sentence frame to solve the problem: ____ **eighths –** ____ **eighths =** ____ **eighths.** | **Number Sense**<br><br>Divide students into pairs and have them do the Emerging Level activity. Then invite them to write a new word problem based on the one they just solved. Provide this sentence frame: **Ann ate ____ slices of pizza and Teresa ate ____ slices of pizza. The pizza had 8 slices. What is the difference in the amount of pizza they ate written as a fraction?** Have pairs choose two numbers to fill in the blanks. Then have them solve the new problem. Once they have finished, have them trade papers with another set of partners to check answers. |

**Teacher Notes:**

NAME _____ DATE _____

# Lesson 3 Guided Writing

## Inquiry/Hands On: Use Models to Subtract Like Fractions

**How do you subtract like fractions using models?**

Use the exercises below to help you build on answering the Essential Question. Write the correct word or phrase on the lines provided.

1. Rewrite the question in your own words.
   See students' work.
   _____

2. What key words do you see in the question?
   like fractions, subtract, models

3. A __unit__ fraction has a numerator of 1.

4. Is the fraction modeled below a unit fraction? __yes__

   $\boxed{\dfrac{1}{10}}$

5. How many fraction tiles are used to model the fraction $\frac{9}{10}$? __nine__

6. How many fraction tiles are used to model the fraction $\frac{2}{10}$? __two__

7. When you subtract $9 - 2$, you take away __2__ from __9__.

8. When you subtract $\frac{9}{10} - \frac{2}{10}$, you take away __2__ unit fractions.

9. Model $\frac{9}{10} - \frac{2}{10}$, find the difference. __$\frac{7}{10}$__

10. How do you subtract like fractions using models?
    Model the subtraction using unit fractions with a denominator that is
    the same as the like fractions. Use the same number of unit fraction
    tiles as the numerator of the subtrahend. Remove the same number of
    unit fraction tiles as the numerator in the minuend fraction.

# Lesson 4 Subtract Like Fractions
## English Learner Instructional Strategy

### Vocabulary Support: Activate Prior Knowledge

Before the lesson, draw a KWL Chart on the board. Write: $\frac{2}{10} + \frac{6}{10} = \frac{8}{10}$.

Say, *This equation includes like fractions. What are like fractions?*

**fractions with the same denominator** Write *like fractions*, along with students' definition, in the chart's K column. Then ask, *When adding like fractions, do we add the numerators or denominators?* **the numerators** *What happens to the denominators?* **They stay the same.** Add these notes to the K column. Then write this in the W column and ask: *How do we subtract like fractions?* Guide students to surmise that, in subtraction, as in addition, the numerators are subtracted, and the denominators stay the same. Then say, *In this lesson we will learn how to subtract like fractions.* Have students help you complete the chart's L column following the lesson.

### English Language Development Leveled Activities

| Emerging Level | Expanding Level | Bridging Level |
|---|---|---|
| **Listen and Identify** <br><br> Write $\frac{5}{8}$ and $\frac{3}{8}$ on the board. Point to the denominator in each fraction and say, *The **denominator** is 8.* Have students repeat chorally. Then circle the denominators in the fractions and say, *The denominators are the **same**. So, these are **like fractions**.* Have students repeat chorally. Now write $\frac{7}{8}$ and $\frac{3}{4}$ on the board. Point to each denominator and say, *This denominator is 8. This denominator is 4.* Then circle the denominators and say, *The denominators are different. So, these are **not** like fractions.* Have students repeat chorally. | **Number Sense** <br><br> Write $\frac{11}{12} - \frac{6}{12} = \frac{\square}{\square}$ on the board. Circle the numerators and say, *Subtract the numerators.* Write 5 as the numerator in the solution. Then circle the denominators and say, *Keep the same denominator.* Write 12 as the denominator in the solution. Then write additional subtraction problems with like fractions and solve them with students' input. | **Act It Out** <br><br> Have each student draw a circle to represent a pizza and then draw lines to divide their pizza into "slices." Next, have students write a real-world word problem which requires the subtraction of like fractions, based on their pizza slices. Finally, tell students to trade pizzas and papers with a partner and model each other's subtraction problem. Have students simplify the difference, if necessary. |

### Teacher Notes:

NAME _____ DATE _____

# Lesson 4 Vocabulary Cognates
## *Subtract Like Fractions*

Use the Glossary to define the math word in English and in Spanish in
the word boxes. Write a sentence using your math word.

| like fractions | fracciones semejantes |
|---|---|
| **Definition**<br>Fractions that have the same denominator. | **Definición**<br>Fracciones que tienen el mismo denominador. |

**My math word sentence:**

Sample answer: $\frac{1}{5}$ and $\frac{2}{5}$ are like fractions.

| simplest form | mínima expresión |
|---|---|
| **Definition**<br>A fraction in which the numerator and denominator have no common factor greater than 1. | **Definición**<br>Fracción en la que el numerador y el denominador no tienen ningún factor común mayor que 1. |

**My math word sentence:**

Sample answer: $\frac{3}{5}$ is the simplest form of $\frac{6}{10}$.

# Lesson 5 Problem-Solving Investigation
# Strategy: Work Backward

## English Learner Instructional Strategy

### Graphic Support: Anchor Chart

Display and have students read aloud this anchor chart while solving Apply the Strategy Exercise 1:

1. **Understand:** The whole has _____ parts. Chloe finished _____ part(s) after dinner. She has _____ part(s) left. I need to find _____.

2. **Plan:** Work backward to solve the problem.

3. **Solve:** [The whole] − [the part finished after dinner] − [the unfinished part] = [the part Chloe finished before dinner].

4. **Check:** Add as follows: [The part finished before dinner] + [the part finished after dinner] + [the unfinished part] = 1 whole, or _____.

As needed, guide students to see that knowing the whole has 6 parts, or equals $\frac{6}{6}$, gives them the key to solving the problem.

### English Language Development Leveled Activities

| Emerging Level | Expanding Level | Bridging Level |
|---|---|---|
| **Word Recognition** | **Listen and Identify** | **Problem Solving** |
| Take several steps forward and say, *I moved forward.* Have students repeat chorally, while mimicking the motion. Then take several steps backward and say, *I moved backward.* Have students repeat chorally, while mimicking the motion. Next, draw a simple maze and label its start and finish. Model working from "Start to Finish" to solve it, and say, *I worked forward.* Then model working from "Finish to Start" to solve it, and say, *I worked backward.* Have students repeat chorally. Now, create a new maze and ask students to model the terms *forward* and *backward.* | Write, then read aloud this word problem: *I have a meeting at work at 4:30. It takes me 15 minutes to get to work. What time should I leave?* Using the demonstration clock, show students that if you start at 4:30 and move backward 15 minutes, the time is 4:15. *So, you will need to leave at 4:15.* Then explain that you found the answer using the work backward strategy. Have students chorally repeat, **work backward strategy**. Create additional time problems and have students solve them using demonstration clocks. | Write, then read aloud this word problem: *The time on Sonia's cell phone is 5:00. She spent $\frac{1}{2}$ hour riding her bicycle, and $2\frac{1}{2}$ hours studying. At what time did Sonia begin riding her bicycle?* Guide students in using the work backward strategy to find the solution. **2:00** Have students write similar time problems. When they have finished, tell them to switch papers with a partner and use the work backward strategy to find the solution to their partner's problem. |

NAME _____ DATE _____

# Lesson 5 Problem-Solving Investigation
## STRATEGY: Work Backward

Work backward to solve each problem.

1. **Chloe** did **some** of her homework **before** dinner.

   **She** did $\frac{2}{6}$ of her homework **after** dinner. She has $\frac{1}{6}$ of her homework **left**.

   What **fraction** of her homework did Chloe do **before** dinner?

   Write in simplest form.

| Understand | Solve |
|---|---|
| I know:<br><br><br>I need to find: | |
| **Plan** | **Check** |
| | |

2. There were **12** goals scored **during** the game.

   **Team A** scored $\frac{8}{12}$ of the goals.

   **Team B** scored **2** goals during the **first half** of the game.

   What **fraction** of the goals did **Team B** score during the **second half** of the game?

   Write in simplest form.

| Understand | Solve |
|---|---|
| I know:<br><br>I need to find: | |
| **Plan** | **Check** |
| | |

# Lesson 6 Add Mixed Numbers
## English Learner Instructional Strategy

### Collaborative Support: Partners Work/Pairs Check

Before the lesson, ask, *What are the different parts of a mixed number?* **a whole number and a fraction** Then write: $1\frac{1}{3} = \frac{4}{3}$. Ask, *Which of these is an improper fraction?* $\frac{4}{3}$ Label the fractions appropriately with the terms *mixed number* and *improper fraction*. *How is an improper fraction different from a mixed number?* **It does not include a whole number. The numerator is larger than the denominator.**

During Independent Practice Exercises 2–10, have students work with a partner. Have one student complete a problem while the other works as coach. Then have students switch roles for the next problem. When they finish, have them team up with another pair to check answers. After both pairs have agreed on the answers, ask them to shake hands and continue working in original pairs on the next two problems.

### English Language Development Leveled Activities

| Emerging Level | Expanding Level | Bridging Level |
|---|---|---|
| **Number Sense** <br><br> Write $3\frac{1}{4} = \frac{13}{4}$ on the board. Point to $3\frac{1}{4}$ and say, *The mixed number is $3\frac{1}{4}$.* Then point to $\frac{13}{4}$ and say, *The improper fraction is $\frac{13}{4}$.* Ask volunteers to write additional examples of mixed numbers and improper fractions on the board. Have students identify them using these sentence frames: **The mixed number is _____. The improper fraction is _____.** | **Act It Out** <br><br> Distribute fraction tiles to students. Then write $2\frac{2}{3}$ on the board. Point to the mixed number and say, *Two and two thirds is a mixed number.* Have students repeat chorally. Then model using fraction tiles to decompose the mixed number. Have students mimic, using their own fraction tiles. Then, beside the mixed number, write: $= 1 + 1 + \frac{1}{3} + \frac{1}{3}$. Finally, write other simple mixed numbers on the board and help students model decomposing them. | **Number Game** <br><br> Tell students they are going to race to decompose a mixed number and find the mixed number's equivalent improper fraction. Then write a mixed number on the board. The first student to correctly decompose the mixed number and identify its improper fraction gets to write the next mixed number for the group to decompose. |

### Multicultural Teacher Tip

If students are using algorithms or methods unfamiliar to you, but they are still attaining the correct solution, encourage them to demonstrate the methods of solving they've learned in their native countries. Model the preferred algorithm or method used in the United States, and point out any similarities to or differences with the student's approach. Have the student demonstrate for others how he or she solves a problem differently than the method being taught. Perhaps have two students solve the same problem side-by-side using two different algorithms.

NAME _____ DATE _____

# Lesson 6 Note Taking
## *Add Mixed Numbers*

Read the question. Write words you need help with and research each word. Use your lesson to write your Cornell notes. Write or draw math examples to explain your thinking. Share your examples with a classmate.

| | |
|---|---|
| **Building on the Essential Question**<br><br>How do you add mixed numbers? | **Notes:**<br><br>A number that has a whole number part and a fraction part is a __mixed__ __number__.<br><br>The mixed number $4\frac{2}{3}$ decomposed into a sum of whole numbers and unit fractions is equal to __1__ + __1__ + __1__ + __1__ + $\frac{1}{3}$ + $\frac{1}{3}$.<br><br>The mixed number __2__ $\frac{1}{3}$ can be decomposed into $1 + 1 + \frac{1}{3}$.<br><br>A fraction with a numerator that is greater than or equal to the denominator is an __improper__ __fraction__. |
| **Words I need help with:**<br>See students' words. | The sum of the whole numbers, $1 + 1 + 1 + 1 + 1 + 1$ is __6__.<br><br>The sum of the unit fractions, $\frac{1}{3} + \frac{1}{3} + \frac{1}{3}$ is $\frac{3}{3}$<br><br>I can use the equivalent fractions of one whole to write an improper fraction as a mixed number.<br><br>$4\frac{2}{3} + 2\frac{1}{3} = 6\frac{3}{3} = $ __7__ |

**My Math Examples:**
See students' examples.

# Lesson 7 Subtract Mixed Numbers

## English Learner Instructional Strategy

### Language Structure Support: Tiered Questions

Before the lesson, write *equivalent*. Underline the *equ* and ask, *What synonym of* **equivalent** *also starts with these letters?* **equal**

During the lesson, use the following tiered questions to help students with the HOT Problems exercise:

**Emerging:** *Let's solve this problem. Should we work forward?* **no** *So, let's work backward. Do we add or subtract?* **add** *What do we add?* $3\frac{1}{3} + 2\frac{1}{3}$ *What is the sum?* $5\frac{2}{3}$

**Expanding/Bridging:** *Should we work forward or backward to solve this problem?* **backward** *What expression should we use?* $3\frac{1}{3} + 2\frac{1}{3}$ *What is the sum?* $5\frac{2}{3}$

### English Language Development Leveled Activities

| Emerging Level | Expanding Level | Bridging Level |
|---|---|---|
| **Word Recognition** | **Building Oral Language** | **Number Game** |
| Write $\frac{11}{5}$ on the board. Say, *I'm going to* **simplify** *the improper fraction.* Have students repeat chorally. Then write this on the board: $= 2\frac{1}{5}$. Point to the mixed number and say, *This is the* **simplest form**. Now ask, *Is this a* **mixed number**? Students should respond, **yes** or show a thumbs-up. Repeat with additional examples of improper fractions that need to be converted to simplest form. | Write this on the board: $\frac{16}{6} = 2\frac{4}{6} = 2\frac{2}{3}$. Point to $2\frac{2}{3}$ and say, *The simplest form is* $2\frac{2}{3}$. Then write additional improper fractions on the board. Have students identify the simplest form of each fraction using this sentence frame: **The simplest form is _____.** | On one index card, write a mixed number. Then create a "match" for this card by writing the mixed number's equivalent improper fraction on another index card. Prepare enough pairs of cards in this manner so that each student will get one card. Then shuffle the cards and distribute them to students. Say, *Find the person who has the mixed number or improper fraction that matches your card.* After all matches have been located, collect all the cards, shuffle them, and redistribute cards, to repeat the activity. |

**Teacher Notes:**

NAME _____ DATE _____

# Lesson 7 Concept Web

*Subtract Mixed Numbers*

Use the concept web to write the equivalent improper fraction of each mixed number.

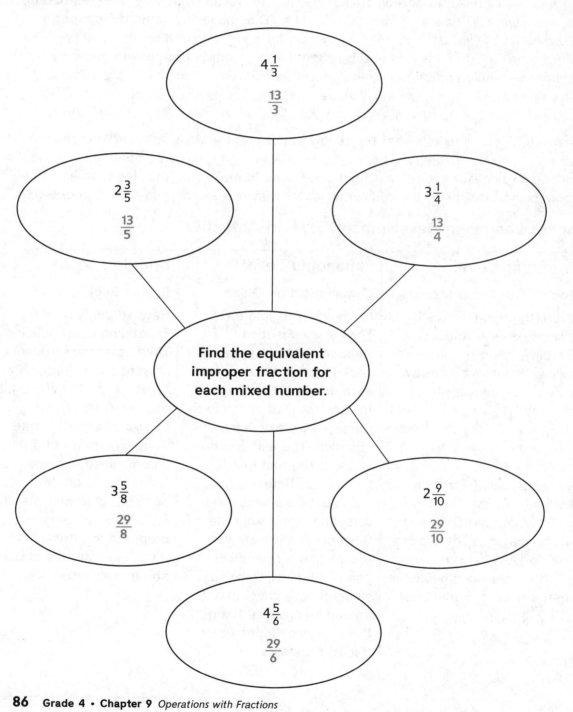

# Lesson 8 Inquiry/Hands On:
# Model Fractions and Multiplication
## *English Learner Instructional Strategy*

### Vocabulary Support: Activate Prior Knowledge

Before the lesson, ask, *Is multiplication the same as repeated addition or repeated subtraction?* **repeated addition** Then have students recall what they have learned about like fractions. Ask, *How are like fractions alike?* **They have the same denominator.** *When we add like fractions, does the numerator change?* **yes** *Does the denominator change?* **no** Tell students that in this lesson they will learn about how to multiply fractions by whole numbers. Say, *Multiplication by a whole number is the same as repeated addition. So, what will change when fractions are multiplied by whole numbers? Will the numerators change?* **yes** *Will the denominators change?* **no**

Finally, have students tell what they know about unit fractions. Ask, *What is the numerator of a unit fraction?* **1** *Do all unit fractions have the same denominator?* **no** Ask students to say examples of unit fractions. Check to make sure that they are accurately pronouncing the /th/ at the ends of words such as *one-fourth, one-fifth, and one-sixth.*

### English Language Development Leveled Activities

| Emerging Level | Expanding Level | Bridging Level |
|---|---|---|
| **Look, Listen, and Identify** Work through Practice It Exercise 3 with students. Simplify the instructions as follows. Point to $\frac{3}{4}$ and say, *This is the product.* Write and say, *We need to: 1) Tell the product's unit fraction. 2) Use the product to complete this problem:* $\frac{3}{4} =$ **multiple × unit fraction.** Remind students that the unit fraction has the same denominator as the product. Then ask, *What is the unit fraction?* **one-fourth** Rewrite the formula: $\frac{3}{4} =$ **multiple** $\times \frac{1}{4}$ Ask, *How many times do we add $\frac{1}{4}$ to get $\frac{3}{4}$?* **three** Rewrite the formula: $\frac{3}{4} = 3 \times \frac{1}{4}$ | **Communication Guide** Divide students into pairs. Then assign Practice It Exercises 3–12. Have partners work together to complete each problem. Give them this communication guide to use as they set up each equation: **The unit fraction is ____. The multiple is ____. ____ times ____ is ____.** Once students have completed their work, have them check their answers against those of another pair of students. Once all students have checked and agreed on answers, bring the class back together to discuss responses. | **Report Back** Have students do the Expanding Level activity. Then have each student present one of his or her responses to the class. Ask each student to write out his or her equation and then explain it using this communication guide: **____ times ____ equals ____. So, ____ is a multiple of ____.** Tell students to be prepared to answer questions from the class about their responses. |

### Teacher Notes:

NAME _____ DATE _____

# Lesson 8 Note Taking

## Inquiry/Hands On: Model Fractions and Multiplication

Read the question. Write words you need help with and research each word. Use your lesson to write your Cornell notes. Write or draw math examples to explain your thinking. Share your examples with a classmate.

| **Building on the Essential Question**<br><br>How do you model the multiplication of fractions? | **Notes:**<br><br>Find the sum. $2 + 2 + 2 =$ _6_<br><br>Find the product. $3 \times 2 =$ _6_<br><br>Multiplication can be thought of as repeated addition.<br><br>Two unit fraction tiles are used to model the fraction $\frac{2}{8}$.<br><br>$\boxed{\frac{1}{8} \quad \frac{1}{8}}$<br><br>How many fraction tiles are used to model $\frac{2}{8} + \frac{2}{8} + \frac{2}{8}$? ___six___<br><br>How many fraction tiles are used to model $3 \times \frac{2}{8}$? <br>_six_<br><br>Find the sum. $\frac{2}{8} + \frac{2}{8} + \frac{2}{8} = \frac{6}{8}$<br><br>Find the product. $3 \times \frac{2}{8} = \frac{6}{8}$ |
|---|---|
| **Words I need help with:**<br>See students' words. | |

**My Math Examples:**
See students' examples.

# Lesson 9 Multiply Fractions by Whole Numbers

## English Learner Instructional Strategy

### Language Structure Support: Sentence Frames

Before the lesson, review the term *repeated addition*. Ask, *What operation is like repeated addition?* **multiplication**

Also, before beginning Math in My World Example 2, point out this sentence: *The product lies between the whole numbers 1 and 2.* Tell students that *lies* is a multiple-meaning word, and discuss its many definitions. Ensure students understand that, in Example 2, *lies* means "is located."

Finally, provide the following sentence frames for students to use as they explain their solutions to the Problem Solving exercises:

**11. The friends do/do not have enough rope because _____.**

**12. Mrs. Raymond will need _____ boxes because _____.**

## English Language Development Leveled Activities

| Emerging Level | Expanding Level | Bridging Level |
|---|---|---|
| **Developing Oral Language** Write $4 \times \frac{5}{6} = 3\frac{1}{3}$ on the board. Circle $3\frac{1}{3}$ and say, *This is the **product**.* Have students repeat chorally. Then draw a number line from 2 to 4 that is divided into increments of thirds. Circle $3\frac{1}{3}$ on the number line and say, *The product is **between** 3 and 4.* Have students repeat chorally. Repeat the activity using different examples of fractions multiplied by whole numbers. | **Number Recognition** Write $8 \times \frac{1}{5} = 1\frac{3}{5}$ on the board. Then draw a number line on the board from 1 to 3 divided into increments of fifths. Label the hash mark for $1\frac{3}{5}$ and say, *The product is **between** 1 and 2.* Have students repeat chorally. Then write additional equations with fractions multiplied by whole numbers. Have students identify the two whole numbers between which the product lies, using this sentence frame: **The product is between _____ and _____.** | **Public Speaking Norms** Assign each student a multiplication problem in which a fraction is multiplied by a whole number. Direct students to find the solution and simplify it as needed. Then divide students into multilingual groups. Have students present their problem to their group and explain how they found the solution. |

## Teacher Notes:

NAME _____ DATE _____

# Lesson 9 Guided Writing

## Multiply Fractions by Whole Numbers

**How do you multiply a fraction by a whole number?**

Use the exercises below to help you build on answering the Essential Question. Write the correct word or phrase on the lines provided.

1. Rewrite the question in your own words.
   <u>See students' work.</u>

2. What key words do you see in the question?
   <u>fraction, whole number, multiply</u>

3. When you <u>decompose</u> a number, you break it into different parts.

4. Decompose the fraction $\frac{3}{8}$ into the sum of a unit fraction.
   <u>$\frac{1}{8} + \frac{1}{8} + \frac{1}{8}$</u>

5. Decompose the fraction $\frac{3}{8}$ into the product of a whole number and a unit fraction.
   <u>$3 \times \frac{1}{8}$</u>

6. Use the Associative Property to rewrite the multiplication.
   $7 \times \frac{3}{8} = 7 \times \left(3 \times \frac{1}{8}\right) = (\underline{\ 7\ } \times \underline{\ 3\ }) \times \frac{1}{8}$

7. Find the product of $7 \times \frac{3}{8}$
   <u>$\frac{21}{8}$ or $2\frac{5}{8}$</u>

8. How do you multiply a fraction by a whole number?
   <u>Use repeated addition, or models. Rewrite the fraction as a product of a whole number and unit fraction. Multiply (using Associative Property) and simplify.</u>

**88** **Grade 4 · Chapter 9** *Operations with Fractions*

# Chapter 10 Fractions and Decimals

## What's the Math in This Chapter?

### Mathematical Practice 2: Reason abstractly and quantitatively.

On the floor use painters or masking tape to create a large number line that is divided into tenths. On 11 pieces of paper, individually label fractions $\frac{1}{10}$ through $\frac{9}{10}$. Make one paper with zero "0". Say, *I want to label this number line using fractions. How many sections are in my number line?* **10** *So what number should be the denominator of each of my fractions?* **10** Place the fraction labels below the number line.

Have a student stand on $\frac{3}{10}$. Ask the other students to identify the fraction he or she is standing on. **three-tenths** Say, *I want to show this number in another way. What decimal could I use to show this number?* **three-tenths** Write $\frac{1}{10} = 0.3$ on the board. Repeat until all fractions and their decimal equivalents have been listed.

Discuss and model that the pronunciation of the fraction in tenths is pronounced the same as the fraction in decimal form. Then discuss how the fractions and decimals are the same value using a base-ten model. Say, *Decimals and fractions can **represent** or show the same number. When they do, the **quantities** are the same.*

Display a chart with Mathematical Practice 2. Restate Mathematical Practice 2 and have students assist in rewriting it as an "I can" statement, for example: **I can understand different representations for the same number.** Post the new "I can" statement.

## Inquiry of the Essential Question:

### How are fractions and decimals related?

Inquiry Activity Target: **Students come to a conclusion that numbers can be represented by different models.**

As an introduction to the chapter, present the Essential Question to students. The inquiry graphic organizer will offer opportunities for students to observe, make inferences, and apply prior knowledge of models representing the Essential Question. As they investigate, encourage students to draw, write, and collaborate with peers to demonstrate their observations and thinking. Then have students present additional questions they may have to a peer to extend discussions.

Regroup students and restate Mathematical Practice 2 and the Essential Question. Pose questions to reflect on what has been learned to guide students in making connections between the Mathematical Practice and the Essential Question.

NAME _____ DATE _____

# Chapter 10 Fractions and Decimals

## Inquiry of the Essential Question:

### How are fractions and decimals related?

Read the Essential Question. Describe your observations (I see...), inferences (I think...), and prior knowledge (I know...) of each math example. Write additional questions you have below. Then share your ideas and questions with a classmate.

$\frac{13}{100}$

| Ones | Tenths | Hundredths |
|:----:|:------:|:----------:|
| 0.   | 1      | 3          |

I see ...

I think...

I know...

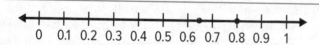

0   0.1  0.2  0.3  0.4  0.5  0.6  0.7  0.8  0.9  1

0.8 is to the right of 0.65. So, 0.8 > 0.65.

I see ...

I think...

I know...

Questions I have...

_____

_____

_____

# Lesson 1 Inquiry/Hands On: Place Value Through Tenths and Hundredths

## English Learner Instructional Strategy

### Language Structure Support: Communication Guide

Before assigning the Apply It Exercises, define words and phrases that may be unfamiliar to students. For example, explain *flower, petals, storm, fall off, left, movie theater, seats,* and *full.* Use photos, illustrations, realia, and demonstrations to reinforce meaning. Further support Emerging Level students by pairing them with Expanding/Bridging Level students to work on the problems. Provide students with this communication guide to help them form a response to Apply It Exercise 15: **I had $2.00. I spent $____ on a candy bar. Now, I have $0.58 left.** Clarify words in the communication guide that may be confusing, such as the multiple-meaning words *bar* and *left.* Discuss reasons why the word problem you have provided is an example of a *real-world* problem: *Spending money is something people do in their real lives; it is realistic for a candy bar to cost between $1.00 and $2.00.*

### English Language Development Leveled Activities

| Emerging Level | Expanding Level | Bridging Level |
|---|---|---|
| **Look, Listen, and Identify** | **Multiple Word Meanings** | **Show What You Know** |
| During the Build It activitiy, point out the fractions $\frac{1}{10}$ and $\frac{1}{100}$. Ask, *Are these unit fractions?* **yes** *How do you know? What part of the fraction tells you, the numerator or the denominator?* **the numerator** Remind students that a unit is always one of something. Then point to the fraction $\frac{1}{100}$. Explain that this fraction represents the smallest monetary unit in dollars: the *cent.* Say, *One cent is the least amount of money you can own in U.S. dollars.* | Write *decimal point* and underline *point.* Explain that *point* is a word with many meanings. Use realia, pictures, and demonstrations to explain some of its more common meanings. Use a pencil to demonstrate the meaning "a sharp end." Say, *This is the point of a pencil.* Use your finger to point at a classroom window, to demonstrate the meaning "to indicate a place, person, or object." Say, *Let's all point to the window.* Finally, show students that point can also mean "a small circular shape," as in the case of a decimal point. | Have each student roll a number cube twice to fill in the last two blanks of this decimal: **0._ _.** Ask them to shade a hundreds model to represent the decimal. Have them label their models using this sentence frame: **My model shows ____ hundredths.** Once they are finished, their decimal should be represented three ways: a model, number form, and written form. Remind students to include a hyphen between the two number words used to name the decimal (for example: twenty-five). Post their work in the classroom. |

### Multicultural Teacher Tip

Because many word problems involve prices and/or determining changes in monetary value, ELs will benefit from an increased understanding of American coins and bills. A chart or other kind of graphic organizer visually comparing coin and bill values and modeling how to write dollars and cents in decimal form would help these students. You may also want to have ELs describe the monetary systems of their native countries. Identifying similarities or differences with the American system can help familiarize students with dollars and cents.

NAME _____ DATE _____

# Lesson 1 Concept Web

## Inquiry/Hands On: Place Value Through Tenths and Hundredths

Use the concept web to write the place value of the 3 in each number.

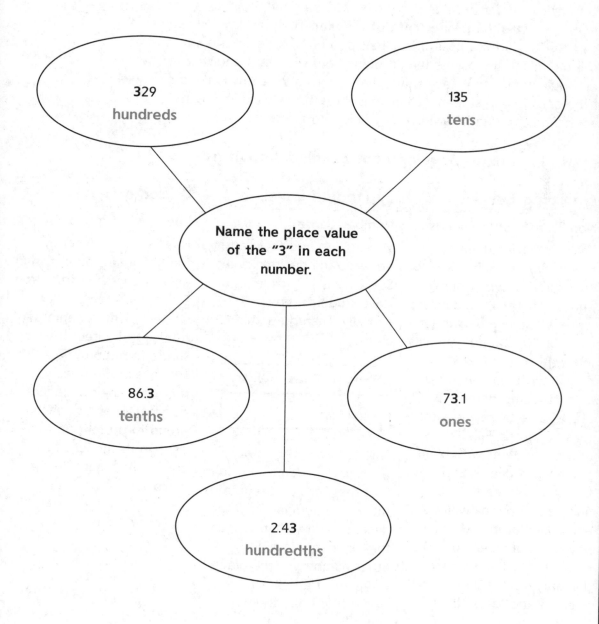

# Lesson 2 Tenths

## English Learner Instructional Strategy

### Sensory Support: Mnemonic Device

Before the lesson, write *decimal* and its Spanish cognate, *decimal*.
Introduce the words, and provide a math example. Utilize other
appropriate translation tools for non-Spanish speaking ELs.

Then review the term *tenth*. Ask, *Why was one tenth compared to a dime
in Lesson 1?* **A dime is one tenth of a dollar.** Then display a place-value
chart that includes a position for tens and ones, a decimal point, and a
position for tenths. Discuss the difference between tens and *tenths*. Then
point out that, in the place-value chart, "tens and tenths make a sandwich
of the ones." Have students chant this three times. Guide students to
understand that tenths divide one whole into ten parts.

### English Language Development Leveled Activities

| Emerging Level | Expanding Level | Bridging Level |
|---|---|---|
| **Non-Transferable Sounds** <br><br> Say and have students repeat words ending with /th/ such as: *with, fourth,* and *math.* Then, randomly write whole numbers in tens (10, 20, 30, ...) and decimals in tenths (0.1, 0.2, 0.3, ...). As you write each number, have students identify its place value by saying either **tens** or **tenths.** Finally, provide students with blank place-value charts. Explain that you will read aloud numbers and they will write each number in their chart. Say these numbers: 0.8, 0.2, 15, 7, 3, and 0.4. Have students display their answers and assess their understanding. | **Number Sense** <br><br> Write 25.4, and draw an arrow to the decimal point. Say, *This is a decimal point.* Have students repeat chorally. Then draw this chart: <br><br> <table><tr><td>**Tens**</td><td>**Ones**</td><td>**Tenths**</td></tr><tr><td>2</td><td>5 .</td><td>4</td></tr></table> <br> Say, *This number is twenty-five **and** four tenths.* Explain that the word *and* separates the whole number from the decimal value. Then repeat the exercise by writing other numbers in the place-value chart and having students identify them chorally. | **Show What You Know** <br><br> Have students work in multilingual groups. Assign each group a decimal from 0.1 to 0.9. Then provide students with an enlarged tenths grid. Direct them to shade the appropriate sections in the grid to represent their assigned decimal. Then ask each group to present and explain its model to the class. Post models in the classroom or hallway. |

### Teacher Notes:

NAME _____ DATE _____

# Lesson 2 Vocabulary Cognates
## *Tenths*

Use the Glossary to define the math word in English and in Spanish in the word boxes. Write a sentence using your math word.

| **decimal** | **decimal** |
|---|---|
| **Definition** | **Definición** |
| A number with one or more digits to the right of the decimal point such as 8.37 or 0.05. | Número con uno o más dígitos a la derecho del punto decimal, como 8.37 o 0.05. |
| **My math word sentence:** | |
| Sample answer: A dollar and twenty cents is represented by the decimal $1.20. | |

| **tenth** | **décima** |
|---|---|
| **Definition** | **Definición** |
| One of ten equal parts, or $\frac{1}{10}$. | Una de diez partes igualeos o $\frac{1}{10}$. |
| **My math word sentence:** | |
| Sample answer: A dime is equal to one tenth of a dollar. | |

# Lesson 3 Hundredths
## English Learner Instructional Strategy

### Language Structure Support: Communication Guide

Before the lesson, review the term *hundredth*. Ask, *Why was one hundredth compared to a penny in Lesson 1?* **A penny is one hundredth of a dollar.**

During the lesson, have students work with a partner to complete Exercises 9–12. Ask them to take turns describing the money pictured using this communication guide:

1. There are _____ quarters. They equal _____ hundredths.

2. There are _____ dimes. They equal _____ hundredths.

3. There is _____ nickel. It equals _____ hundredths.

4. There [is/are] _____ [penny/pennies]. [It/They] equal(s) _____ hundredths.

5. All the coins equal _____ hundredths, or 0._____.

## English Language Development Leveled Activities

| Emerging Level | Expanding Level | Bridging Level |
|---|---|---|
| **Building Oral Language** | **Number Sense** | **Act It Out** |
| On the board, randomly write whole numbers in hundreds (100, 200, 300, …) and decimals in hundredths (0.01, 0.02, 0.03, …). As you write each number, have students identify it's place value by saying either **hundreds** or **hundredths**. Then provide students with blank place-value charts. Explain that you will read aloud numbers and they will write each number in their chart. Read aloud these numbers: 0.68, 0.04, 0.93. Discuss answers as a group. | Draw a place-value chart, and write the number 430.12. Point to the 4 and say, *The number in the hundreds place is four.* Have students repeat chorally. Then point to the 2 and say, *The number in the hundredths place is two.* Have students repeat chorally. Finally, have pairs of students practice writing similar numbers, such as 529.37 and 145.51. Ask them to identify numbers in the hundreds and hundredths place using these sentence frames: **The number in the hundreds place is _____. The number in the hundredths place is _____.** | Have students work in pairs. Ask one student to write a decimal to the hundredths place and the other student to use coins to model the decimal. (For example, if the decimal is 0.59, it should be represented using two quarters, one nickel, and four pennies.) Then ask both students to name the decimal. Finally, have students switch roles to continue the activity. Provide support as needed. |

**Teacher Notes:**

NAME _____ DATE _____

# Lesson 3 Vocabulary Definition Map
*Hundredths*

Use the definition map to write a description and list characteristics about the vocabulary word or phrase. Write or draw math examples. Share your examples with a classmate.

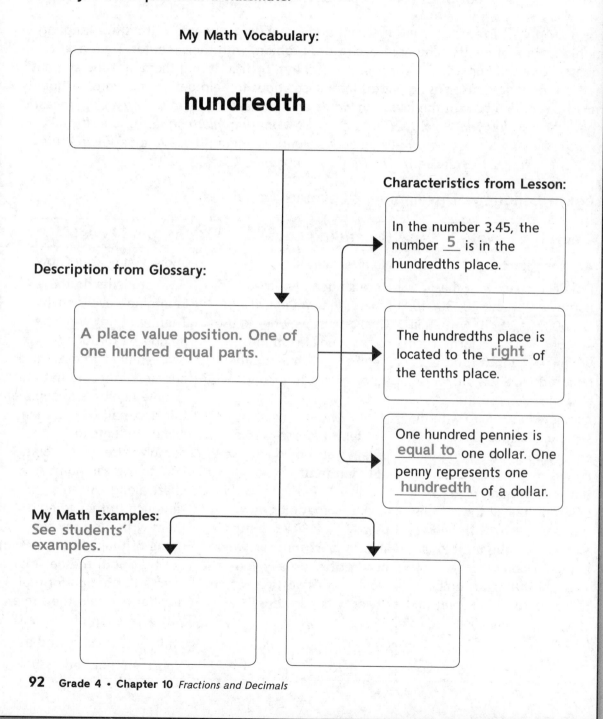

My Math Vocabulary:

**hundredth**

Characteristics from Lesson:

In the number 3.45, the number _5_ is in the hundredths place.

Description from Glossary:

The hundredths place is located to the _right_ of the tenths place.

A place value position. One of one hundred equal parts.

One hundred pennies is _equal to_ one dollar. One penny represents one _hundredth_ of a dollar.

My Math Examples:
See students' examples.

**92** Grade 4 · **Chapter 10** *Fractions and Decimals*

# Lesson 4 Inquiry/Hands On: Model Decimals and Fractions

## English Learner Instructional Strategy

### Collaborative Support: Native Language Mentors

After students have completed the Practice It Exercises, have each student present one of the models he or she shaded and describe it using these sentence frames: **This model shows _____ parts out of _____ equal parts. The fraction is _____ tenths/hundredths.**

For the Apply It Exercises, pair Emerging and Expanding students with native language mentors. Have students echo read each word problem with their mentor. Then ask mentors to clarify unfamiliar words in the problem by translating them into the student's native language and/or using photos and illustrations to help define the words. Students may have difficulty with the following words and phrases: *race track, snowed, gymnastics competition, nine years old, fourth grade class, donating, clothing items, charity, sweaters, jeans.* Students may also need help understanding these language structures: *If _____, then _____* (Exercise 17), *neither _____ nor _____* (Exercise 20).

## English Language Development Leveled Activities

| Emerging Level | Expanding Level | Bridging Level |
|---|---|---|
| **Anchor Chart** Display a chart with the numbers 1–19 and 20, 30, 40, 50, 60, 70, 80, 90 written in number and word form. Then write: **_____ tenths and _____ hundredths.** Show students how they can use the number chart along with the sentence frames for help with writing and saying the names of fractions. Explain that *one* through *nine* can be used to fill in the "tenths" frame and *eleven* through *nineteen* can be used to fill in the "hundredths" frame. Then show how to combine numbers, such as *twenty* and *five*, to fill in the "hundredths" frame. | **Number Sense** Have students do the Emerging Level activity. Then discuss why not to use certain numbers in the "tenths" and "hundredths" frames. For example, ask, *Should we use ten in the "tenths" frame?* **no** *Why not?* **Ten tenths is one whole; usually you would say "one," not "ten tenths."** *Should we use twenty in the hundredths frame?* **sometimes, not usually** *Why?* **Two tenths is the simplest form of twenty hundredths; usually you would say "two tenths," not "twenty hundredths."** | **Show What You Know** Have students do the Emerging-Level activity. Then fill a hat with the numbers 1 through 9. Ask each student to draw two numbers from the hat and use the numbers to complete this decimal: **0._ _.** Then invite students to demonstrate for the class how to use the number chart along with the sentence frames to help them write and say the name of their decimal. Point out that both the decimal form and fraction form of a number are said the same way. (For example, both 0.14 and $\frac{14}{100}$ are expressed as "fourteen hundredths.") |

## Teacher Notes:

NAME _____ DATE _____

# Lesson 4 Guided Writing

## Inquiry/Hands On: Model Decimals and Fractions

**How do you model decimals and fractions?**

**Use the exercises below to help you build on answering the Essential Question. Write the correct word or phrase on the lines provided.**

1. Rewrite the question in your own words.
   <u>See students' work.</u>

2. What key words do you see in the question?
   <u>decimal, fraction, model</u>

3. The word form of the fraction $\frac{3}{10}$ is *three tenths*.

4. Identify the decimal represented by the model shown. Write the decimal in the place-value chart.

 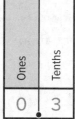

5. The word form of the fraction $\frac{30}{100}$ is *thirty hundredths*.

6. Identify the decimal represented by the model shown. Write the decimal in the place-value chart.

 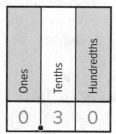

7. The decimals 0.3 and 0.30 are <u>equivalent</u>. The fractions $\frac{3}{10}$ and $\frac{30}{100}$ are <u>equivalent</u>.

8. How do you model decimals and fractions?
   <u>Shade parts of a grid that has ten or a hundred equal parts.</u>
   <u>Find the decimal or fraction on a number line.</u>

Grade 4 • **Chapter 10** *Fractions and Decimals* **93**

# Lesson 5 Decimals and Fractions

## English Learner Instructional Strategy

### Collaborative Support: Turn and Talk

Before the lesson, write this on the board: 0.9. Then ask, *What fraction means the same as 0.9?* Have students think about how to answer your question and then turn and talk to a neighbor about their ideas. If necessary, remind students of how they used models in Lesson 4 to change decimals to fractions. Then call on a volunteer to answer your question by saying the fraction and writing it on the board. $\frac{9}{10}$

During the lesson, provide these sentence frames to help students form their responses to Exercise 16: **A decimal with _____ digit is equal to a fraction with a denominator of 10. A decimal with _____ digit(s) is equal to a fraction with a denominator of 100.**

### English Language Development Leveled Activities

| Emerging Level | Expanding Level | Bridging Level |
|---|---|---|
| **Listen and Identify** Write this equation on the board: $0.6 = \frac{6}{10}$. Say, *When the decimal is **tenths**, the denominator is 10.* Then write this equation: $0.33 = \frac{33}{100}$. Say, *When the decimal is **hundredths**, the denominator is 100.* Finally, write other decimals on the board. Have students identify the decimal place by saying either, **tenths/hundredths**. Then have students identify the denominator by saying, **tenths/hundredths** for the equivalent fraction. | **Number Sense** Write $\frac{2}{10}$ and $\frac{2}{100}$ on the board. Explain the pattern of zeros, using tenths and hundredths grid models. Say, *Add a zero to both the numerator and denominator of the fraction $\frac{2}{10}$ to create the equivalent fraction $\frac{2}{100}$.* Then write another fraction that has 10 as its denominator. Ask students to write an equivalent fraction with a denominator of 100 on write-on/wipe-off boards. Then have a volunteer approach the board and write the equivalent fraction. Discuss answers, and then repeat the activity using a new fraction. | **Synthesis** Distribute a copy of Work Mat 5: Tenths and Hundredths Models to each student. Have students shade in a portion of each tenth grid to represent 6 different tenths decimals. Then have students trade grids with a partner. Direct students to shade the hundredths grids to show an equivalent decimal for each tenths grid. Then have them write equivalent decimals and equivalent fractions for each pair of grids. Finally, have partners check each other's work. |

**Teacher Notes:**

NAME _____ DATE _____

# Lesson 5 Four-Square Vocabulary

*Decimals and Fractions*

Write the definition for each math word. Write what each word means in your own words. Draw or write examples that show each math word meaning. Then write your own sentences using the words.

| Definition | My Own Words |
|---|---|
| A number that uses place value, numbers, and a decimal point to show part of a whole. | See students' examples. |

**decimal**

| My Examples | My Sentence |
|---|---|
| Sample answer: 0.1, 0.05, 0.25 | Sample sentence: The value of a quarter can be written using the decimal 0.25. |

| Definition | My Own Words |
|---|---|
| A number that represents part of a whole or part of a set. | See students' examples. |

**fraction**

| My Examples | My Sentence |
|---|---|
| Sample answer: $\frac{5}{100}$, $\frac{3}{10}$, $\frac{1}{4}$ | Sample sentence: A quarter has a value that is $\frac{1}{4}$ of a dollar. |

# Lesson 6 Use Place Value and Models to Add

## English Learner Instructional Strategy

### Language Structure Support: Report Back

Before the lesson, review the term like fractions. Write: $\frac{7}{10}$, $\frac{51}{100}$. Then ask, *Are these like fractions?* **No** *Why?* **They have different denominators.** Discuss how to make $\frac{7}{10}$ and $\frac{51}{100}$ like fractions, guiding students through the process of changing $\frac{7}{10}$ to $\frac{70}{100}$.

During the lesson, provide these sentence frames to help students report back:

1. First, I made like fractions by changing the fraction with a denominator of 10 into an equivalent fraction with a denominator of _____. [100]

2. Then I added the _____ [like] fractions to get a sum of _____.

3. Finally, I wrote the sum _____ as the decimal _____.

### English Language Development Leveled Activities

| Emerging Level | Expanding Level | Bridging Level |
|---|---|---|
| **Number Sense**<br>Write: $\frac{5}{10} + \frac{42}{100}$. Point to and name each denominator. Say, *These are **unlike** fractions, so we **cannot** add them.* Then place a zero after the numerator and denominator of the first fraction so that it becomes $\frac{50}{100}$. Point to and name each denominator again. Say, *These are **like** fractions, so we **can** add them.* Then solve the equation: $\frac{50}{100} + \frac{42}{100} = \frac{92}{100}$, or 0.92. Point to the sum as a fraction and say, *ninety-two hundredths.* Then point to the sum as a decimal and say, *ninety-two hundredths.* Have students repeat. | **Show What You Know**<br>Write $\frac{3}{10} + \frac{45}{100}$ on the board. Ask a volunteer to demonstrate changing the expression into one that has like fractions with denominators of 100. (The student should change $\frac{3}{10}$ to $\frac{30}{100}$.) Then write an equals sign after the expression and solve with student input: $\frac{30}{100} + \frac{45}{100} = \frac{75}{100}$. Have another student come to the board and write the sum as a decimal: **0.75**. Finally, repeat the exercise, using similar addition problems. | **Number Game**<br>Have students work in pairs. Ask one partner to write a fraction with 10 as a denominator and the other partner to write a fraction with 100 as the denominator. Then have partners reveal their fractions to one another at the same time and race to find the sum of their fractions. Tell students to write the sum as both a decimal and a fraction with a denominator of 100. The first student to complete the addition problem wins. Repeat as time allows. |

## Teacher Notes:

NAME _____ DATE _____

# Lesson 6 Note Taking
## Use Place Value and Models to Add

Read the question. Write words you need help with and research each word. Use your lesson to write your Cornell notes. Write or draw math examples to explain your thinking. Share your examples with a classmate.

| | |
|---|---|
| **Building on the Essential Question** | **Notes:** |
| How can you use place value and models to add? | The fraction $\frac{2}{10}$ is represented in the decimal model below. |
| | 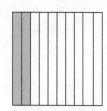 |
| | The fraction $\frac{20}{100}$ is represented in the decimal model below. |
| | 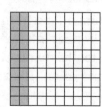 |
| **Words I need help with:** | These two fractions are <u>equivalent</u>. |
| See students' words. | Before adding $\frac{2}{10} + \frac{45}{100}$, write the fractions as like fractions. |
| | Like fractions are fractions that have the same <u>denominator</u>. |
| | $\frac{2}{10} + \frac{45}{100} = \frac{20}{100} + \frac{45}{100} = \frac{65}{100}$ |
| | The sum, written in decimal form, is <u>0.65</u>. |

**My Math Examples:**

See students' examples.

Grade 4 • **Chapter 10** *Fractions and Decimals* **95**

# Lesson 7 Compare and Order Decimals

## English Learner Instructional Strategy

### Collaborative Support: Numbered Heads Together

For Exercises 16–19, divide students into groups of four. Assign each student a number from one to four. Then ask students to solve Exercise 16. Have groups discuss the decimal values, agree on the correct order, and ensure everyone in their group understands and can give the answer. Then call out a number, 1 to 4, randomly. Have students assigned to that number raise their hands and, when called on, answer for their team. Continue this routine for Exercises 17–19.

For HOT Problem Exercise 26, encourage students to draw a number line that includes marks for 0.36 through 0.48, to help them visualize and determine the mid-point.

### English Language Development Leveled Activities

| Emerging Level | Expanding Level | Bridging Level |
|---|---|---|
| **Background Knowledge** | **Number Sense** | **Number Recognition** |
| Draw a number line on the board from 0 to 1, and divide it into tenths. Label the tenths marks 0.1, 0.2, 0.3, and so on. Then draw an identical number line. This time, do not label the tenths marks in order. For example, from left to right, write 0.3, 0.7, 0.2, and so on. Then point to the first number line and say, *The decimals are in order.* Have students repeat chorally. Next, point to the second number line and say, *The decimals are not in order.* Have students repeat chorally. Finally, erase the first number line, and ask students to fix the order of the second one. | Draw a number line from 0 to 1, and divide it into tenths. Label the tenths marks 0.1, 0.2, 0.3, and so on. Then write 0.7 > 0.4. Circle 0.4 and then 0.7 on the number line, and say, *Seven tenths is greater than four tenths.* Have students repeat chorally. Then write 0.4 < 0.7. Again, reference 0.4 and 0.7 on the number line, and say, *Four tenths is less than seven tenths.* Have students repeat chorally. Finally, write other pairs of decimals. Have students compare them using this sentence frame: _____ **is greater than/less than** _____. | Write a variety of decimals on individual index cards, making some of the decimals tenths and some hundredths. Then give one card to each student. Divide students into four groups. Then have the students in each group work together to arrange the cards in order *least* to *greatest*. Finally, ask students in each group to present their ordered decimals. For a challenge, have all the groups combine their decimal cards and place them in order *least* to *greatest*. |

**Teacher Notes:**

NAME _____ DATE _____

# Lesson 7 Guided Writing

## Compare and Order Decimals

**How do you compare and order decimals?**

Use the exercises below to help you build on answering the Essential
Question. Write the correct word or phrase on the lines provided.

1. Rewrite the question in your own words.
   See students' work.
   _____

2. What key words do you see in the question?
   decimal, compare, order

3. When you move to the **right** on the number line below, the whole numbers
   increase in value.

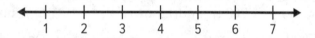

4. When you move to the **left** on the number line above, the whole numbers
   decrease in value.

5. When you move to the **right** on the number line below, the decimals
   increase in value.

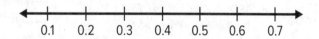

6. When you move to the **left** on the number line above, the decimals
   decrease in value.

7. How do you compare and order decimals?
   Graph the decimals on a number line. The decimal that appears the
   most or farthest to the right on the number line is the greatest.

# Lesson 8 Problem-Solving Investigation Strategy: Extra or Missing Information

## English Learner Instructional Strategy

### Vocabulary Support: Cognates

Have a bilingual aide review the lesson and Problem-Solving Exercises using the online Spanish Student Edition with students. For non-Spanish speaking ELs, provide an appropriate translation tool for the aide to utilize.

Have students compare the English Student Edition lesson to the Spanish Student Edition lesson and identify cognates. Ask students to write a list and then add the following cognates to the cognate chart: *family/familia, day/dia, plan/planea, part/parte, mile/milla, information/información, time/tiempo, determine/determina, problem/problemas, action/acción, comedies/comedias, adventure/aventuras, basketball/básquetbol, vacation/vacaciones, number/noemero, favorite/favorito, vanilla/vainilla, chocolate/chocolate, cereal/cereales.*

### English Language Development Leveled Activities

| Emerging Level | Expanding Level | Bridging Level |
| --- | --- | --- |
| **Act It Out** | **Listen and Identify** | **Synthesis** |
| Demonstrate the meaning of *extra.* Have a student put on several of the other students' jackets. Then say, *[Student name] is wearing **extra** coats. Some coats are **not needed.*** Have students repeat with you, ***Extra* means "not needed."** Then demonstrate the meaning of *missing.* Distribute books to three students, and make a point of showing that you do not have a book for a fourth and fifth student. Say, *[Student names] have missing books. More books are **needed.*** Have students repeat with you, ***Missing* means "needed."** | <table><tr><td></td><td>Ben Saved</td><td>Maya Saved</td><td>Riko Saved</td></tr><tr><td>Jan.</td><td>$12.45</td><td>$0.00</td><td>$9.00</td></tr><tr><td>Feb.</td><td>$11.25</td><td>$14.00</td><td>$9.00</td></tr></table> Draw the above chart. Then tell students you want to know how much Maya and Riko saved in January and February. Ask, *Does the chart have **extra** information or **missing** information?* **extra** Next, say you want to know how much Ben and Riko saved January through April. Ask, *Does the chart have **extra** information or **missing** information?* **Missing.** Repeat with similar questions. | Have students write a real-world word problem with extra or missing information. Refer students to problems in the lesson for examples, and encourage them to incorporate fractions. Then have students exchange papers with a classmate and solve each other's problem. Finally, have students meet in multilingual groups to present their problems and explain the solutions. Ensure that students correctly identify and describe the extra or missing information in each word problem. |

**Teacher Notes:**

NAME _____ DATE _____

# Lesson 8 Problem-Solving Investigation
## STRATEGY: Extra or Missing Information

**Determine if there is extra or missing information to solve each problem. Then solve if possible.**

1. There are **100 movies** at the store.
   $\frac{30}{100}$ are **action** movies, $\frac{50}{100}$ are **comedies**, and $\frac{20}{100}$ are **adventure** movies.

   What **part** of the movies are **action or comedies**?

| Understand | Solve |
|---|---|
| I know: | |
| I need to find: | |
| **Plan** | **Check** |
| The extra or missing information is: | |

2. In a basketball game, the **red team** scored $\frac{3}{10}$ of the baskets during the **first half** and $\frac{4}{10}$ of the baskets during the **second half**.
   The blue team had 10 players.
   How many baskets did the **red team** score during the **first half and second half** of the game?

| Understand | Solve |
|---|---|
| I know: | |
| I need to find: | |
| **Plan** | **Check** |
| The extra or missing information is: | |

**Grade 4 · Chapter 10** *Fractions and Decimals* **97**

# Chapter 11 Customary Measurement
## What's the Math in This Chapter?

### Mathematical Practice 6: Attend to precision.

Have three student volunteers come to the front of the room. Hand each of them a one cup measuring cup filled with water. Then distribute a piece of string, an inch ruler, and a balance scale (one item each) to the students and say, *Please measure.* Allow time for the volunteers to think and attempt to measure the water or the cup. Ask observing students if they have any ideas that will help the volunteers measure. Discuss ideas and observations. The discussion goal should be for students to recognize that you didn't tell them what to measure or how to measure.

Say, *I forgot to tell you what to measure and what unit of measurement to use. I should have been more precise. We need to measure the capacity.* Display a clear 2-cup measuring cup and say, *We need to use a different tool to measure capacity. The items I gave you measure weight and length, not capacity.* Pour the water from one measuring cup into your 2-cup measuring cup and say, *The capacity is 8 ounces.* Discuss that it is important to be precise when measuring. It is also important to specify the unit of measure.

Display a chart with Mathematical Practice 6. Restate Mathematical Practice 6 and have students assist in rewriting it as an "I can" statement, for example: **I can carefully represent measurements.** Post the new "I can" statement in the classroom.

## *Inquiry of the Essential Question:*

### Why do we convert measurements?

Inquiry Activity Target: **Students come to a conclusion that measuring carefully is very important.**

As an introduction to the chapter, present the Essential Question to students. The inquiry graphic organizer will offer opportunities for students to observe, make inferences, and apply prior knowledge of measurement representing the Essential Question. As they investigate, encourage students to draw, write, and collaborate with peers to demonstrate their observations and thinking. Then have students present additional questions they may have to a peer to extend discussions.

Regroup students and restate Mathematical Practice 6 and the Essential Question. Pose questions to reflect on what has been learned to guide students in making connections between the Mathematical Practice and the Essential Question.

NAME _____ DATE _____

# Chapter 11 Customary Measurement

## *Inquiry of the Essential Question:*

**Why do we convert measurements?**

Read the Essential Question. Describe your observations (I see...), inferences (I think...), and prior knowledge (I know...) of each math example. Write additional questions you have below. Then share your ideas and questions with a classmate.

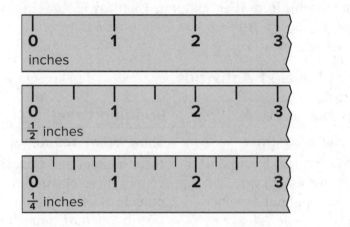

I see ...

I think...

I know...

| seconds (s) | minutes (min) | (s, min) |
|-------------|---------------|-----------|
| 60 | 1 | (60, 1) |
| 120 | 2 | (120, 2) |
| 180 | 3 | (180, 3) |
| 240 | 4 | (240, 4) |

I see ...

I think...

I know...

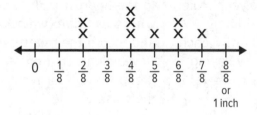

I see ...

I think...

I know...

Questions I have...

_____

_____

_____

# Lesson 1 Customary Units of Length

## English Learner Instructional Strategy

### Vocabulary Support: Making Connections

Before the lesson, write *yard* and its Spanish cognate, *yarda*. Introduce the term, and provide a math example. Utilize other appropriate translation tools for non-Spanish speaking ELs. Also, use photos to explain non-math meanings for *yard*. Some students may be more familiar with the word *vara*, meaning 3 feet.

Then write and explain *customary system*. Underline the word *custom* in *customary* and say, *Something that is **customary** is done or used as part of a **custom**.* Explain that a custom is a habit or practice and that many cultures have their own customs, or ways of doing things. Invite students to share examples of customs from their cultures. Then say, *There are even different customs for measuring things. In this lesson we will learn about systems of measurement used in the United States.*

### English Language Development Leveled Activities

| Emerging Level | Expanding Level | Bridging Level |
|---|---|---|
| **Multiple Meaning Words** | **Word Recognition** | **Show What You Know** |
| Point to your foot and say, *This is a **foot**. There are toes on this foot.* Then hold up a 12-inch ruler and say, *This is a **foot**.* Explain that this kind of foot is used for measuring. Point to each inch mark on the ruler as you count to 12. Then say, *Twelve inches equals one foot.* Have students chorally repeat. Next, show a picture of a house with a yard. Point to the yard and say, *This is a **yard**. There is grass in this yard.* Then hold up a yardstick and say, *This is a **yard**.* Show students that the yardstick is the same length as three 12-inch rulers. Say, *One yard equals three feet.* Have students chorally repeat. | Hold a paper clip against a ruler to show that it measures about one inch. Say, *This is about one **inch**.* Have students repeat chorally. Next, display a 12-inch ruler and say, *This is one **foot**.* Have students repeat chorally. Then show a yardstick and say, *This is one **yard**.* Have students repeat chorally. Finally, display the paper clip, ruler, and yardstick randomly. Have students identify each unit of measure by saying **inch**, **foot**, or **yard**. | Have multilingual groups collect three classroom objects of varying lengths. Distribute chart paper to each group. Then ask each group member to estimate each object's length to the nearest inch. Have students trace their objects onto the chart paper and record their estimates. Finally, have them measure each object to the nearest inch to see whose estimate was closest and record the measurement. Have volunteers from each group share their results with the class. |

## Multicultural Teacher Tip

As the metric system is the standard throughout most parts of the world, ELs will most likely be more familiar with units of metric measurement than they will be with standard units. Students who have worked only with the metric system in the past will be more familiar with partial amounts written as decimals, not fractions.

NAME _____ DATE _____

# Lesson 1 Multiple Meaning Word
## *Customary Units of Length*

Complete the four-square chart to review the multiple meaning words
*foot* **and** *yard*.

| Everyday Use | Math Use in a Sentence |
|---|---|
| Sample answer: A person has a foot at the bottom of his or her leg. | Sample sentence: The twig measured one foot in length. |

**foot (ft)**

| Math Use | Example From This Lesson |
|---|---|
| A customary unit for measuring length. Plural is feet. | Sample answer: 12 inches is equal to 1 foot |

| Everyday Use | Math Use in a Sentence |
|---|---|
| Sample answer: An outdoor area next to a building. | Sample sentence: The desk measured one yard in length. |

**yard (yd)**

| Math Use | Example From This Lesson |
|---|---|
| A customary unit for measuring length equal to 3 feet or 36 inches. | Sample answer: 3 feet is equal to 1 yard |

**Write the correct numbers on each line to complete the sentence.**

Since 1 yard = <u>3</u> feet and 1 foot = <u>12</u> inches, then you know 3 feet = <u>36</u> inches and 1 yard = <u>36</u> inches.

# Lesson 2 Convert Customary Units of Length

## English Learner Instructional Strategy

### Graphic Support: Word Web

Before the lesson, write *convert* in the center of a Word Web. Tell students that a word that means the same as *convert* is *change*. Then have students help you fill out the Word Web with different examples of converting. For example, money from one country can be converted to the money of another country. Then write *conversion* and explain that it means "the act of converting." Discuss ways students have already used conversion in math, such as in problems where fractions with different denominators were changed into like fractions. Then write *conversion table*, refer them to an example in the lesson, and briefly explain how it will be used to convert units of measurement.

### English Language Development Leveled Activities

| Emerging Level | Expanding Level | Bridging Level |
|---|---|---|
| **Background Knowledge** | **Number Sense** | **Show What You Know** |
| Write: 1 foot. Say, *We will convert to inches. Convert means "change."* Then display a ruler and say, *One foot is equivalent to 12 inches.* Write: 1 foot = 12 inches. Create a conversion table to show the measurement equivalencies. Finally, show how to convert 2 feet and then 3 feet into inches and complete the conversion table. Encourage student input as you demonstrate the conversions. | Write: 96 inches. Then explain how to convert inches to feet. Write then say, *Inches to feet, divide by 12.* Have students chant 3 times chorally. Then solve 96 ÷ 12 = 8 and say, *96 inches is equivalent to 8 feet.* Now write: 5 feet. Explain how to convert feet to inches. Write then say, *Feet to inches, multiply by 12.* Have students chant 3 times chorally. Solve 5 × 12 and say, *5 feet is equivalent to 60 inches.* Guide students in solving other conversions between inches and feet. Before solving, have students say the appropriate chant aloud to identify the correct operation. | Have students work in multilingual groups to create a conversion table on chart paper that includes inches, feet, and yards. Encourage them to refer to tables from the Guided and Independent Practice exercises for examples. Assign the following measurements for conversion: 36 inches, 72 inches, 3 yards, 12 feet, 5 yards. Then have each group present and explain its table to the class. |

**Teacher Notes:**

NAME _____ DATE _____

# Lesson 2 Vocabulary Cognates

## Convert Customary Units of Length

Use the Glossary to define the math word in English and in Spanish in
the word boxes. Write a sentence using your math word.

| **convert** | **convertir** |
|---|---|
| **Definition** | **Definición** |
| To change one unit to another. | Cambiar de una unidad a otra. |
| **My math word sentence:** | |
| Sample answer: 3 feet can be converted to 1 yard. 6 feet can be converted to 2 yards. | |

| **mile (mi)** | **milla (mi)** |
|---|---|
| **Definition** | **Definición** |
| A customary unit of measure for length | Unidad usual de longitud. |
| **My math word sentence:** | |
| Sample answer: 1 mile = 5,280 feet | |

# Lesson 3 Customary Units of Capacity

## English Learner Instructional Strategy

### Vocabulary Support: Frontload Academic Vocabulary

Before the lesson, refer to the New Vocabulary with cognates and write them on the cognate chart. Introduce the words, and provide realia to support understanding. Discuss the term *cup*. Explain that within the context of this lesson, a cup is a unit for measuring capacity. Discuss other meanings of *cup* and provide photos of various kinds of cups. Show them a real tea cup, which is comparable in capacity to (and possibly the basis for) the one-cup measure used in cooking.

### English Language Development Leveled Activities

| Emerging Level | Expanding Level | Bridging Level |
|---|---|---|
| **Word Knowledge** | **Developing Oral Language** | **Public Speaking Norms** |
| Gather containers with capacities of one ounce, one cup, one pint, one quart, and one gallon, and label each container with its capacity. Then display each container, identify its unit of capacity, and have students repeat the term chorally. Finally, ask a volunteer to display each container randomly while covering its label. Have the group shout out the container's capacity. Then have the volunteer reveal the container's label. Repeat with other volunteers until students show understanding. | Label five containers with capacities of one ounce, one cup, one pint, one quart, and one gallon. Gather many other containers with a variety of capacities and shapes. As you display each unlabeled container, have students identify the most reasonable unit of capacity using the labeled containers as a comparison. Provide the following sentence frame: **That container's capacity looks equal to one ____ (ounce/cup/pint/quart/ gallon) because ____ .** | Gather containers with a variety of capacities and shapes. Distribute 7 to 10 containers to multilingual groups. Ask groups to discuss and then group the containers according to similar capacities. Have volunteers from each group explain and justify their groupings utilizing lesson vocabulary. |

**Teacher Notes:**

NAME _____ DATE _____

# Lesson 3 Vocabulary Definition Map
## *Customary Units of Capacity*

Use the definition map to write a description and list characteristics about the vocabulary word or phrase. Write or draw math examples. Share your examples with a classmate.

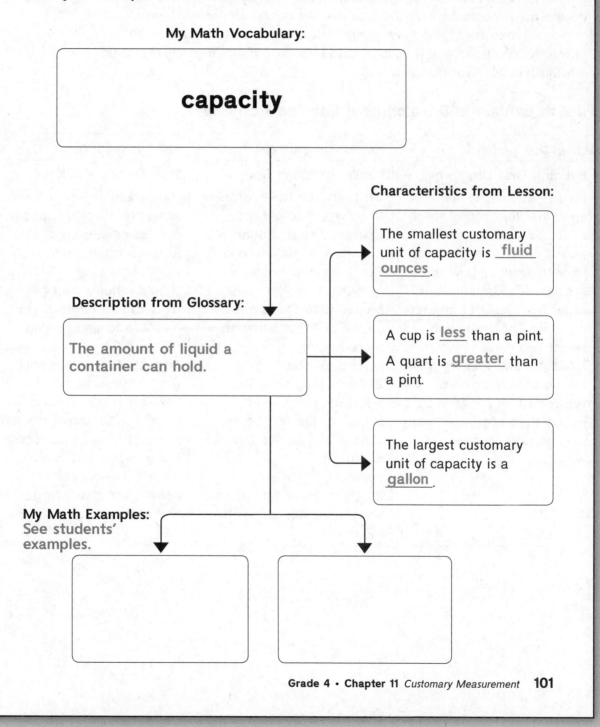

**My Math Vocabulary:**

**capacity**

**Characteristics from Lesson:**

The smallest customary unit of capacity is <u>fluid ounces</u>.

**Description from Glossary:**

The amount of liquid a container can hold.

A cup is <u>less</u> than a pint.
A quart is <u>greater</u> than a pint.

The largest customary unit of capacity is a <u>gallon</u>.

**My Math Examples:**
See students' examples.

**Grade 4 · Chapter 11** *Customary Measurement* **101**

# Lesson 4 Convert Customary Units of Capacity

## English Learner Instructional Strategy

### Sensory Support: Realia and Mnemonics

Before the lesson, demonstrate the meaning of "To change a larger unit to a smaller unit, multiply." Display a gallon jug of water and 4 empty quart containers. Model converting the gallon into quarts by pouring the water into the quart containers. Write: 1 gallon = 4 quarts. Underneath write: $1 \times 4 = 4$. Then say, *I have more smaller units than the larger unit, so I multiply.* Write more and multiply and underline the *m* in each. Continue with Math in My World Example 1.

### English Language Development Leveled Activities

| Emerging Level | Expanding Level | Bridging Level |
|---|---|---|
| **Building Oral Language** | **Public Speaking Norms** | **Show What You Know** |
| Display and name one container for each of these capacities: ounce, cup, pint, quart, gallon. Ask students to tell/show you how to arrange the containers in order from least to greatest capacity. Then randomly select two units of capacity. Have students identify their relationship using these sentence frames: **A(n) _____ is less than a(n) _____. A(n) _____ is greater than a(n) _____.** | Write this on the board: 1 cup = 8 fluid ounces, 1 pint = 2 cups, 1 quart = 2 pints, 1 gallon = 4 quarts. Say, *I want to know the number of ounces in 2 cups.* Model using the information on the board to arrive at this equation for finding the number of ounces in 2 cups: $2 \times 8 = 16$ ounces. Then ask students to find the following: the number of cups in 2 pints, the number of pints in 2 quarts, and the number of quarts in 2 gallons. Have volunteers share their solutions with the class. | Have multilingual partners create real-world capacity conversion word problems for each other. Remind students that the solution should require their partner to change a larger unit of measure to smaller unit of measure. Then encourage each student to present their original problem and their partner's solution to another pair. Direct students to explain how he or she knew to multiply when making the capacity conversion from a larger unit to a smaller unit. |

**Teacher Notes:**

NAME _____ DATE _____

# Lesson 4 Vocabulary Chart
## *Convert Customary Units of Capacity*

Use the three-column chart to organize the review vocabulary in this lesson. Write the word in Spanish. Then write the correct terms to complete each definition.

| English | Spanish | Definition |
|---|---|---|
| **capacity** | capacidad | The amount of <u>liquid</u> a container can hold. |
| **convert** | convertir | To <u>change</u> one unit to another. |
| **is equal to (=)** | es igual a (=) | Having the <u>same</u> value. The = sign is used to show two numbers or expressions are <u>equal</u>. |
| **is greater than (>)** | es mayor que (>) | An <u>inequality</u> relationship showing that the number on the <u>left</u> side of the symbol is <u>greater</u> than the number on the right side. |
| **is less than (<)** | es menor que (<) | An <u>inequality</u> relationship showing that the number on the <u>left</u> side of the symbol is <u>less</u> than the number on the right side. |

# Lesson 5 Customary Units of Weight
## English Learner Instructional Strategy

### Vocabulary Support: Build Background Knowledge

Before the lesson, write: *ounce (oz), pound (lb)*. Help students remember the atypical abbreviations by explaining their origins. Discuss that *oz* is based on the old Italian word *onza*, which means "ounce." Write *onza* and underline its *o* and *z*. Discuss that onza is also the Spanish cognate for *ounce*. Then discuss that *lb* is based on the ancient Roman term *libra pondo*, which means "pound weight." Write *libra pondo* on the board and underline its *l* and *b*. Have students write this information in their math journals.

### English Language Development Leveled Activities

| Emerging Level | Expanding Level | Bridging Level |
|---|---|---|
| **Word Knowledge** | **Show What You Know** | **Public Speaking Norms** |
| Display a bottle that contains a liquid, such as water. Say, *Fluid ounces measure capacity*. Shake the bottle or sprinkle a little of the water out to illustrate the meaning of fluid. Then display an object such as a notepad. Say, *Ounces measure the weight of solids*. Demonstrate weighing the notepad in a bucket balance. Finally, display or show pictures of various solid objects and bottled liquids. Have students say, **fluid ounces** or **ounces** to identify the appropriate unit of measure for each item. | Collect ten items, some of which are more than one pound and some of which are less. Include smaller, heavy items such as a block of wood, as well as larger, lighter items such as an empty box. Then draw a T-chart on the board with the headings Ounces and Pounds. Show students an item from the collection, and have them tell whether they think the item's weight should be measured in ounces or pounds. List student's answers in the chart. Then have students weigh each item in a bucket balance or on scales to check their answers. | Provide multilingual pairs or groups with images of objects that vary in size and weight. (Images can be cut from a magazine or printed from an online source.) Then have students estimate the weight of each object. Finally, ask each pair or group to present their estimates to the class and explain their reasoning. |

### Teacher Notes:

NAME _____ DATE _____

# Lesson 5 Concept Web
## *Customary Units of Weight*

Use the concept web to write the best unit of weight for each object.
The first one is done for you.

> **Word Bank**
>
> ounces                          pounds                          tons
> (A slice of bread weighs        (A football weighs             (A buffalo weighs
> about 1 oz.)                    about 1 lb.)                   about 1 T.)

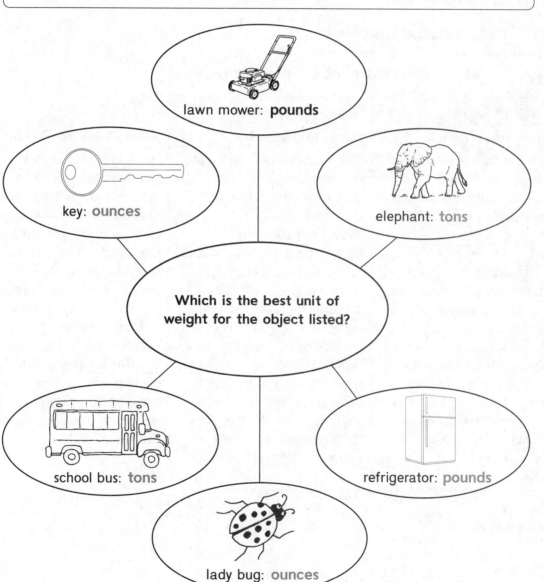

lawn mower: **pounds**

key: ounces

elephant: tons

Which is the best unit of
weight for the object listed?

school bus: tons

refrigerator: pounds

lady bug: ounces

Grade 4 • **Chapter 11** *Customary Measurement*  **103**

# Lesson 6 Convert Customary Units of Weight

## English Learner Instructional Strategy

### Vocabulary Support: Activate Prior Knowledge

Prompt students to recall why, when converting larger units to smaller units, they should multiply. **It takes more of the smaller units than the larger units to measure the same capacity.** Then write the words: *ton, pound, ounce.* Review these terms using them in various ways in the following sentence frame: **I will convert _____ to _____. Should I multiply?** Students should answer **yes** or **no** and then explain their response using this sentence frame:

**A _____ is [larger/smaller] than a _____.**

### English Language Development Leveled Activities

| Emerging Level | Expanding Level | Bridging Level |
|---|---|---|
| **Visual Connections** | **Listen and Identify** | **Show What You Know** |
| Write tons in very large letters then to the right of that write pounds in small letters. Underneath write: 1 ton = 2,000 pounds. Then say, *Larger to smaller, multiply.* Have students chant chorally 3 times. Next, write ounces in even smaller letters to the right of pounds. Underneath write: 1 pound = 16 ounces. Have students repeat the chant. Model solving, 2 T = _____ lb, 3 lb = _____ oz. Then have students choose 3 My Homework exercises to solve on their own. | Pair students and distribute 0–5 number cubes and write-on/wipe-off boards. Have students take turns rolling the cube to determine a number of tons to convert to pounds. For example if 4 is rolled by one student, then the partner converts 4 tons into 8,000 pounds by multiplying 4 by 2,000. Have students switch roles for each round. After several rounds, have students then roll the cube to determine a number of pounds to convert to ounces. | Have students work in multilingual groups to create a conversion table on chart paper that includes tons, pounds, and ounces. Encourage them to refer to tables from the Guided and Independent Practice exercises for examples. Assign the following measurements for conversion: 64,000 ounces, 2,000 pounds, and 3 tons. Then have each group present and explain its table to the class. |

**Teacher Notes:**

NAME _____ DATE _____

# Lesson 6 Guided Writing
## *Convert Customary Units of Weight*

**How do you convert customary units of weight?**

**Use the exercises below to help you build on answering the Essential Question. Write the correct word or phrase on the lines provided.**

1. Rewrite the question in your own words.
   See students' work.

2. What key words do you see in the question?
   convert, customary, units, weight

3. Which is greater, one ton or one pound?
   1 ton

4. Which is greater, one ounce or one pound?
   1 pound

5. To convert from a larger unit to a smaller unit, use the multiplication
   operation.

6. When measuring with different units, it will take more of a smaller unit than of
   the larger unit to reach the same capacity.

7. 1 pound equals 16 ounces. 2 pounds equals 32 ounces.

8. 1 ton equals 2,000 pounds. 3 tons equals 6,000 pounds.

9. How do you convert customary units of weight?
   Determine which customary equivalency will be used (16 ounces = 1 pound;
   1 ton = 2,000 pounds) and then use that fact and multiplication to convert
   the larger unit to the smaller unit.

# Lesson 7 Convert Units of Time
## English Learner Instructional Strategy

### Vocabulary Support: Cognates

Before the lesson, write *seconds, minute, hour,* and *day* and their Spanish cognates respectively, *segundos, minuto, hora,* and *día* on a cognate chart. Introduce the math terms, and provide visual examples such as a demonstration analog clock to support understanding. Utilize other appropriate translation tools for non-Spanish speaking ELs.

Review the meaning of *time interval*. Draw two points on the board, labeling one "Start" and the other "Finish." Then draw a line between the two points. Label the points with the start and finish times for lunch at school. Discuss how many minutes students have to eat lunch. Point to the line and say, *This length of time between the start and finish is an interval. The interval for lunchtime is ____ minutes.*

### English Language Development Leveled Activities

| Emerging Level | Expanding Level | Bridging Level |
|---|---|---|
| **Word Knowledge** | **Developing Oral Language** | **Act It Out** |
| Point to and name the second hand, minute hand, and hour hand on a demonstration clock. Have students repeat chorally each unit of time after you say it. Then randomly point to each of the hands on the clock. Have students identify the appropriate unit of time by saying: **second/minute/ hour.** | Show pictures of children engaged in activities that are familiar to students, such as playing soccer, brushing teeth, or eating breakfast. Ask students how long it takes them to do each activity, and have them respond using this sentence frame: **It takes ____ [minutes/hours] to ____.** | Model how to use and read a stopwatch to time an activity. Provide students with a list of activities, such as: *write your name in cursive, name the months of the year in order, tie a shoe,* or *make a paper airplane.* Have students estimate how long they think it will take to perform each activity. Then have partners take turns using a stopwatch to time each other completing each of the activities. Have students compare estimates to actual times. |

**Teacher Notes:**

NAME _____ DATE _____

# Lesson 7 Note Taking

## Convert Units of Time

Read the question. Write words you need help with and research each word. Use your lesson to write your Cornell notes. Write or draw math examples to explain your thinking. Share your examples with a classmate.

| Building on the Essential Question | Notes: |
|---|---|
| **How do you convert units of time?** | Years, months, weeks, days, hours, minutes, and seconds are all different units of ___time___. |
| | • One hour is ___greater___ than one minute. |
| | • One second is ___less___ than one minute. |
| | • One day is ___greater___ than one hour. |
| | • One week is ___greater___ than one day. |
| | • One month is ___less___ than one year. |
| | To convert from a larger unit to a smaller unit, use the ___multiplication___ operation. |
| **Words I need help with:** | 1 day equals ___24___ hours. |
| See students' words. | $(1 \times 2) = 2$ |
| | 2 days equals ___48___ hours. |
| | 1 week equals ___7___ days. |
| | $(1 \times 3) = 3$ |
| | 3 weeks equals ___21___ days. |
| | To find the number of minutes in 3 hours, multiply the number of minutes in 1 hour by the number ___3___. |
| | 1 hour equals ___60___ minutes. 3 hours equals ___180___ minutes. |

**My Math Examples:**

See students' examples.

# Lesson 8 Display Measurement Data in a Line Plot

## English Learner Instructional Strategy

### Vocabulary Support: Build Background Knowledge

Before the lesson, write *data* and its Spanish cognate, *datos* on a cognate chart. Introduce the term, and provide graphic examples to support understanding. Utilize other appropriate translation tools for non-Spanish speaking ELs.

Explain the term *line plot*. Discuss multiple meanings for *plot,* including its literary definition. Then iterate that, within the context of this lesson, the *plot* in *line plot* refers to marks plotted, or placed, on a line graph. Have students preview the line plot used during Math in My World Example 1. Then briefly discuss how the line plot and tally chart correlate. Ensure students understand that, in this example, the marks in both charts stand for the number of bugs of a given length.

### English Language Development Leveled Activities

| Emerging Level | Expanding Level | Bridging Level |
|---|---|---|
| **Synthesis** | **Act It Out** | **Making Connections** |
| **Number of Pets** | Cut ten sentence strips at various lengths between 1 to 5 inches. Distribute 10 strips to each pair of students. Have pairs measure each strip to the nearest inch and record the data in a tally chart. Assist students with labeling. Then have pairs create a line plot and record the information from the tally chart to the line plot. Ask questions as you gesture to the corresponding data such as: *Which length is most/ least frequent? How long are the strips altogether?* | Have multilingual groups create a tally chart and line plot based on the number of vowels in their group members' first names. Once groups have completed their tally charts, have them represent the data in a line plot. Then invite students to present their work to the class. |

Number of Pets tally chart:

| | |
|---|---|
| 1 | ~~||||~~ ~~||||~~ |
| 2 | ~~||||~~ | |
| 3 | ||| |

Draw the tally chart above, and then draw a blank line plot. Invite students to come to the board to plot the correct number of Xs over each number on the line plot. Ask questions that students can answer with a gesture, such as: *Point to the **most** frequent number of pets. Point to the **least** frequent number of pets. Which number has 3 Xs plotted?*

### Teacher Notes:

NAME _____ DATE _____

# Lesson 8 Vocabulary Definition Map
## *Display Measurement Data in a Line Plot*

Use the definition map to write a description and list characteristics about the vocabulary word or phrase. Write or draw math examples. Share your examples with a classmate.

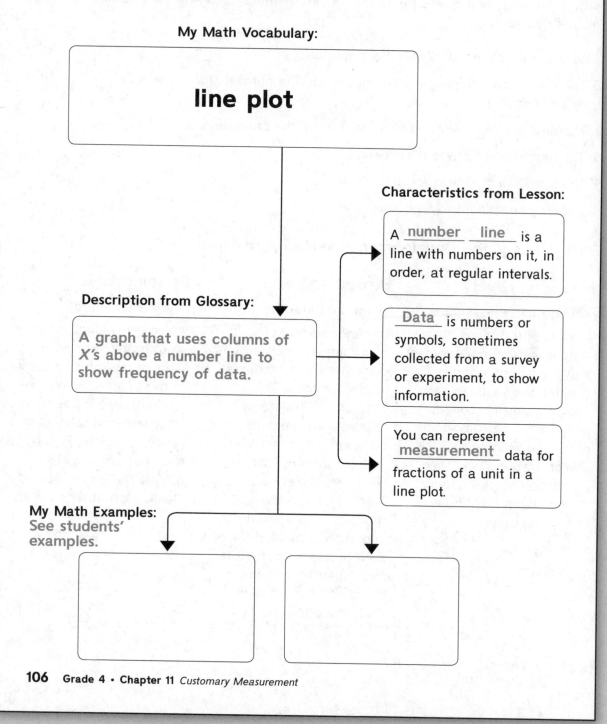

My Math Vocabulary:

**line plot**

Characteristics from Lesson:

A __number__ __line__ is a line with numbers on it, in order, at regular intervals.

__Data__ is numbers or symbols, sometimes collected from a survey or experiment, to show information.

You can represent __measurement__ data for fractions of a unit in a line plot.

Description from Glossary:

**A graph that uses columns of *X*'s above a number line to show frequency of data.**

My Math Examples:
See students' examples.

# Lesson 9 Solve Measurement Problems

## English Learner Instructional Strategy

### Collaborative Support: Round the Table

Copy Exercises 3 and 4 onto separate sheets of paper. Place students into multilingual groups, and distribute both problems to each group. Have students work jointly on each problem by passing the paper around the table for each member to provide input. Direct each member of the group to write with a different color to ensure all students participate in solving the problem. Provide directions:

1. Read the problem aloud as a group and discuss.

2. The first student highlights all signal words and phrases that show which operations to use.

3. The next student circles what to do to solve the problem.

4. The next student solves the problem.

5. The last student checks for reasonableness.

6. Choose one student to report back.

### English Language Development Leveled Activities

| Emerging Level | Expanding Level | Bridging Level |
|---|---|---|
| **Background Knowledge** Review with students the various units of measurement they have used in this chapter. Show students a ruler, a yardstick, containers of various capacities, items of varying weights, and the demonstration clock. Then, as you point to each item, have students identify its unit of measure. | **Listen and Identify** Display a variety of measurement tools and containers students have used throughout the chapter. Use the items to pose measurement problems, such as: *How many more fluid ounces does a 1-pint container hold than a 1-cup container?* Have students identify the operation(s) needed to solve each problem. Then write each operation on the board and solve it with student input. | **Think-Pair-Share** Ask each student to write a word problem based on the units of measure addressed in this chapter. Then have students switch papers with a partner and solve each other's problem. Finally, have partners meet to discuss the problems and check each other's work. |

### Teacher Notes:

NAME _____ DATE _____

# Lesson 9 Note Taking

## *Solve Measurement Problems*

Read the question. Write words you need help with and research each word. Use your lesson to write your Cornell notes. Write or draw math examples to explain your thinking. Share your examples with a classmate.

| **Building on the Essential Question** | **Notes:** |
|---|---|
| How do you solve measurement problems in the customary system? | When solving measurement problems, all values need to be the _____same_____ unit ____before____ performing operations. |
| | There are ___60___ minutes in 1 hour. |
| | 1 hour – 15 minutes = ? <br> ___60___ minutes – 15 minutes = ___45___ minutes |
| | There are ___16___ ounces in 1 pound. |
| | 3 pounds + 10 ounces = ? <br> ( ___16___ × 3) ounces + 10 ounces = ? <br> ___48___ ounces + 10 ounces = ___58___ ounces |
| **Words I need help with:** <br> See students' words. | When a problem asks for the **difference** in units, use the operation of __subtraction__ . |
| | When a problem asks for the **total** number of units, use the operation of __addition__ . |

**My Math Examples:**

See students' examples.

# Lesson 10 Problem-Solving Investigation
## Strategy: Guess, Check, and Revise
### English Learner Instructional Strategy

## Vocabulary Support: Review Basic Vocabulary

Review English terms from the Problem Solving exercises which may be unfamiliar, such as: *elk, mountains, fruit punch, lemonade, lives, water plants, watering can, aquarium, gravel, fish, home, museum, seal, download, songs, digital music player, stunt person, building roof, skydiver, plane.*

Display labeled drawings, photos or realia to clarify word meaning in a prominent location for students to reference during the lesson.

## English Language Development Leveled Activities

| Emerging Level | Expanding Level | Bridging Level |
|---|---|---|
| **Word Knowledge** | **Developing Oral Language** | **Show What You Know** |
| Show students a jar filled with marbles or coins, and have them guess how many items are in the jar. Say to each student, *Guess how many.* Have students use this sentence frame to respond: **I guess _____.** Make a guess yourself that is obviously too high. Then pull out a handful of the items and count them. Ask, *Does anyone want to revise his or her guess?* Explain that *revise* means "change." Model revising your original guess, then allow students to make their revisions. Have students count the number of items to see whose guess was closest. | Work through the Learn the Strategy example with students. Ask them to summarize the steps of the problem-solving strategy. Then guide students, as needed, with prompts such as the following: *The first step is _____. Next, we _____. Then we _____.* Make sure students understand that the process of guessing, checking, and revising can be repeated several times. | Pair bridging students with emerging/expanding students. Assign one of the Review the Strategies exercises. Direct the bridging student to read the problem aloud and have the partner echo read. Have pairs create a graphic organizer on chart paper that shows the sequence of steps they used to solve their exercise. Ask a volunteer from each group to present their graphic organizer. |

**Teacher Notes:**

NAME _____ DATE _____

# Lesson 10 Problem-Solving Investigation
## STRATEGY: Guess, Check, and Revise

Guess, check and revise to solve each problem.

1. **Max** was on vacation <u>twice</u> as long as **Jared** and <u>half</u> as long as **Wesley**. The **boys** were on vacation a <u>total</u> of <u>3 weeks</u>. How many <u>days</u> was <u>each</u> boy on vacation?

| Understand | Solve |
|---|---|
| I know:<br><br>I need to find: | |

| Name | Days |
|---|---|
| Max | 6 |
| Jared | 3 |
| Wesley | 12 |
| Total | 21 |

| Plan | Check |
|---|---|
| Find how many days = 1 week. | |

2. Anu drinks <u>2 cups</u> of water each day. Jan drinks <u>twice</u> as much water as **Anu**. How many <u>fluid ounces</u> does **Jan** drink?

| Understand | Solve |
|---|---|
| I know:<br><br>I need to find: | |

| Plan | Check |
|---|---|
| Find how many fluid ounces = 1 cup. | |

# Chapter 12 Metric Measurement

## What's the Math in This Chapter?

### Mathematical Practice 6: Attend to precision.

Say, *Find me something in the classroom that weighs about 10.* Pretend to be busy as students are looking for items and do not provide any clarification on 10 "what". Collect the objects that students find and discuss how they are alike or different. Students will likely collect items that are fairly large as they may be thinking about 10 pounds.

Display a pen that weighs about 10 grams. Say, *I asked you to find something that weighs about 10 but I forgot to tell you the unit of measurement. I should have been more precise. I should have said 10 grams. Because I was not precise, some of you found items that were too large.* Give students a chance to find items that weigh 10 grams. Compare the items collected to the first group of items. Discuss that stating a unit of measure is helpful and a way to attend to precision in math.

Display a chart with Mathematical Practice 6. Restate Mathematical Practice 6 and have students assist in rewriting it as an "I can" statement, for example: **I can share my measurements by including the units of measure I used.** Have students draw or write examples of measurement units. Post examples and new "I can" statement in the classroom.

## Inquiry of the Essential Question:

### How can conversion of measurements help me solve real-world problems?

Inquiry Activity Target: **Students come to a conclusion that including units of measurement is critical.**

As an introduction to the chapter, present the Essential Question to students. The inquiry graphic organizer will offer opportunities for students to observe, make inferences, and apply prior knowledge of measurement representing the Essential Question. As they investigate, encourage students to draw, write, and collaborate with peers to demonstrate their observations and thinking. Then have students present additional questions they may have to a peer to extend discussions.

Regroup students and restate Mathematical Practice 6 and the Essential Question. Pose questions to reflect on what has been learned to guide students in making connections between the Mathematical Practice and the Essential Question.

NAME _____ DATE _____

# Chapter 12 Metric Measurement

## Inquiry of the Essential Question:

**How can conversion of measurements help me solve real-world problems?**

Read the Essential Question. Describe your observations (I see...),
inferences (I think...), and prior knowledge (I know...) of each math
example. Write additional questions you have below. Then share
your ideas and questions with a classmate.

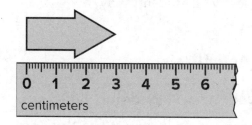

I see ...

I think...

I know...

The arrow is about _____ centimeters long.

I see ...

I think...

I know...

Mass of cat: 5 grams or 5 kilograms?

THINK: 5 grams would have the same mass as
about 5 pennies.

Conversion Table

| Kilograms (kg) | grams (g) | (kg, g) |
|---|---|---|
| 12 | 12,000 | (12,12,000) |
| 14 | 14,000 | (14,14,000) |
| 16 | 16,000 | (16,16,000) |
| 18 | 18,000 | (18,18,000) |

I see ...

I think...

I know...

Questions I have...

_____

_____

_____

**Grade 4 • Chapter 12** *Metric Measurement* **109**

# Lesson 1 Metric Units of Length
## English Learner Instructional Strategy

### Vocabulary Support: Cognates

Before the lesson, write the New Vocabulary and their Spanish cognates on the cognate chart. Introduce the words, and provide realia to support understanding. Utilize appropriate translation tools for non-Spanish speaking ELs. Point out that *meter* is a multiple-meaning word. Discuss definitions for *meter*, as related to the content areas of poetry, music, and math, and show students photos of postage and parking meters. Have students assist in making a word web for *meter*. Then have students transcribe the word web into their math journals. Many EL students may be familiar with the metric system.

### English Language Development Leveled Activities

| Emerging Level | Expanding Level | Bridging Level |
|---|---|---|
| **Developing Oral Language** | **Making Connections** | **Show What You Know** |
| Display a meterstick and say, *This is a meter.* Have students repeat chorally. Next, point out a centimeter increment on the meterstick and say, *This is a centimeter.* Have students repeat chorally. Then, on the board, write: centi $= \frac{1}{100}$. Using the meterstick, show students that there are 100 centimeters in 1 meter. Then have pairs of students measure each other's height in centimeters. Encourage students to tell you and/or a peer their height, using this sentence frame: **I am _____ centimeters tall.** | Using a metric ruler and meterstick, show students a millimeter, a centimeter, and a meter. Name each unit of measure as you point to it, and have students repeat chorally. Then display pictures of items that represent a wide range of sizes, such as a pencil tip, a book, and a swimming pool. Ask students to write the most appropriate metric unit for measuring each item's length—a millimeter, a centimeter, or a meter. Then have them display their answers. | Prepare index cards ahead of time that are labeled with different measurements to take, such as: *width of pencil eraser, length of desktop,* and *height of door.* Give each student one card. Tell students to estimate the length of the object on their card and then measure the object to check their estimates. (Make sure all the card items name objects that can be found in the classroom.) Finally, have students work together to arrange their measures from *least* to *greatest.* |

### Teacher Notes:

NAME _____ DATE _____

# Lesson 1 Vocabulary Chart
## *Metric Units of Length*

Use the three-column chart to organize the vocabulary in this lesson. Write the word in Spanish. Then write the correct terms to complete each definition.

| English | Spanish | Definition |
|---------|---------|------------|
| **centimeter (cm)** | centímetro (cm) | A metric unit for measuring <u>length</u>.<br><br><u>100</u> centimeters = 1 meter |
| **kilometer (km)** | kílometro (km) | A metric unit for measuring <u>length</u>.<br><br>1 kilometer = <u>1,000</u> meters |
| **meter (m)** | metro (m) | A metric unit for measuring <u>length</u>. |
| **metric system (SI)** | sistema métrico (SI) | The <u>decimal</u> system of measurement. Includes units *such* as *meter,* <u>gram</u>, and *liter*. |
| **millimeter (mm)** | milímetro (mm) | A metric unit for measuring <u>length</u>.<br><br><u>1,000</u> millimeters = 1 meter |

# Lesson 2 Metric Units of Capacity
## English Learner Instructional Strategy

### Vocabulary Support: Modeled Talk

Before the lesson, write *capacity, milliliter,* and *liter* and their Spanish cognates, *capacidad, mililitro,* and *litro,* respectively on a cognate chart. Introduce the words and provide realia (eyedropper and liter bottle) to support understanding. Write: milli $= \frac{1}{1000}$. Discuss and partially model that it would take 1,000 milliliters to fill the liter bottle. Then discuss multiple meanings for *capacity.* For example, in addition to meaning "the amount of liquid a container can hold," it can also mean "ability." Use *capacity* in a sentence that demonstrates this meaning as you physically act it out.

During the lesson, have students complete Exercises 4–9 on their own, and then turn and talk to a neighbor to discuss the reasoning behind their responses.

### English Language Development Leveled Activities

| Emerging Level | Expanding Level | Bridging Level |
|---|---|---|
| **Act It Out** | **Developing Oral Language** | **Public Speaking Norms** |
| Distribute a small cup of water, a plastic spoon and an eyedropper labeled with 1 milliliter to each student. Have students fill the eye dropper to the 1 milliliter mark. Then have students put the water into the spoon drop-by-drop as they count each drop. Students should discover there are about 20 drops of water in a milliliter. Then have students determine about how many milliliters it takes to fill the spoon to capacity. Have students write/draw their observations in their math journals using the terms: *milliliters* and *capacity.* | Gather photos of items with a variety of capacities and shapes, and display them for students. Point to a photo and ask, *Would you use an eyedropper which is about 1 milliliter to fill this object to capacity, or would you rather use a liter bottle?* Prompt students to first answer appropriately, **eyedropper/liter bottle.** Then prompt them to answer with the unit of measure, **milliliter/liter.** Repeat with all the photos. | Gather containers with a variety of capacities and shapes. Distribute a variety of containers to multilingual groups. Direct groups to discuss the unit of capacity they would use to measure the capacity of each container. Have volunteers from each group explain their reasoning for choosing either liter or milliliter. |

### Teacher Notes:

NAME _____ DATE _____

# Lesson 2 Vocabulary Cognates

## Metric Units of Capacity

Use the Glossary to define the math word in English and in Spanish in the word boxes. Write a sentence using your math word.

| liter (L) | litro (L) |
|---|---|
| **Definition** | **Definición** |
| A metric unit for measuring volume or capacity. | Unidad métrica para medir el volumen o la capacidad. |
| **My math word sentence:** ||
| Sample answer: 1 liter = 1,000 milliliters ||

| milliliter (mL) | mililitro (mL) |
|---|---|
| **Definition** | **Definición** |
| A metric unit for measuring capacity. | Unidad métrica para medir la capacidad. |
| **My math word sentence:** ||
| Sample answer: 1,000 milliliters = 1 liter ||

Grade 4 • **Chapter 12** *Metric Measurement* **111**

# Lesson 3 Metric Units of Mass
## *English Learner Instructional Strategy*

### Vocabulary Support: Frontload Academic Vocabulary

Before the lesson, write the New Vocabulary and their Spanish cognates on a cognate chart. Introduce the words, and provide realia to support understanding. Utilize appropriate translation tools for non- Spanish speaking ELs.

Point out the words *matter* and *gravity* in the definition and discuss these multiple-meaning words. Ensure students understand what the terms mean within the context of this lesson. Have students create a word web for each term in their math journals, along with notes and pictures that help with remembering their multiple meanings.

### English Language Development Leveled Activities

| Emerging Level | Expanding Level | Bridging Level |
|---|---|---|
| **Listen and Identify** | **Building Oral Language** | **Public Speaking Norms** |
| Display a penny and say, *The penny's mass is about one gram.* Have students repeat chorally. Then hold up a book and say, *The book's mass is about 1 kilogram.* Have students repeat chorally. Finally, give each student a penny and a book. Randomly say *gram* or *kilogram*. Then have students chorally repeat the unit of measure you named and hold up the appropriate object. | Collect ten items, some with a mass more than one kilogram and some with a mass that is less. Then draw a T-chart with the headings Grams and Kilograms. Show students an item from the collection, and ask whether they think its mass should be measured in grams or kilograms. Write the item's name in the chart, under the measure they suggest. Then continue with the remaining items. Finally, have students hold each item and say whether its name should move to a different column of the chart. Have students explain their reasoning. | Provide multilingual pairs or groups of students with images of several common household objects that vary in size and mass. (Images can be cut from a magazine or printed from an online source.) Have students estimate the mass of each object. Then have each pair or group present their estimates to the class and explain their reasoning. |

### Teacher Notes:

NAME _____  DATE _____

# Lesson 3 Concept Web
## *Metric Units of Mass*

Use the concept web to write the best unit of mass to use for each
object. The first one is done for you.

> ### Word Bank
>
> grams                                      kilograms
>
> (A paperclip weighs about 1 g.)    (A loaf of bread weighs about 1 kg.)

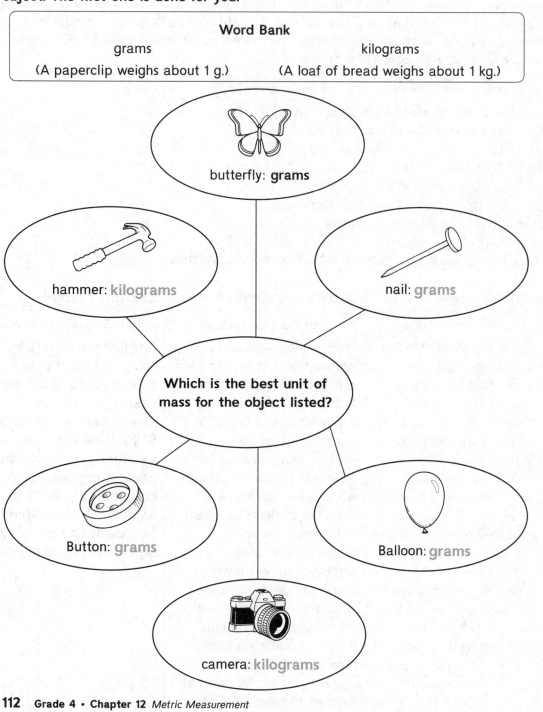

butterfly: **grams**

hammer: kilograms

nail: grams

Which is the best unit of
mass for the object listed?

Button: grams

Balloon: grams

camera: kilograms

# Lesson 4 Problem-Solving Investigation Strategy: Make an Organized List

## English Learner Instructional Strategy

### Collaborative Support: Round the Table

Make copies of Exercises 6 and 7 on separate sheets of paper. Place students into multilingual groups of 4 or 5, and distribute one problem to each group. Have students work jointly by passing their problem around the table for each group member to provide input. Direct each group member to write with a different color to ensure all students participate in solving the problem.

Direct students to follow these participation guidelines:

1. Read the problem aloud as a group, and discuss.
2. One student underlines what they know.
3. The next student circles what to find.
4. The next student writes a plan.
5. The next student solves the problem.
6. The last student checks for reasonableness.
7. Choose one student to present the solution.

### English Language Development Leveled Activities

| Emerging Level | Expanding Level | Bridging Level |
|---|---|---|
| **Background Knowledge** | **Building Oral Language** | **Public Speaking Norms** |
| Point out any lists posted in the classroom, such as a roster of students' names or a poster showing the classroom rules. As you point to each example, say, *This is a list*. Have students repeat chorally. Then show a yellow, a red, and a blue color tile. Say, *I will make a list of possible combinations*. Demonstrate making a list that shows the three possible color combinations for the tiles: *blue/red, blue/yellow, red/yellow*. Then point to one of the tiles. Ask, *Is this a list?* **No** Point to your list on the board. Ask, *Is this a list?* **Yes** | Explain that organizing information in a list can help with problem solving. Draw a T-chart with the headings Sandwiches and Drinks. In the Sandwiches column, write: *cheese, peanut butter, turkey*. In the Drinks column, write: *water, milk, juice*. Then tell students you want to find out how many sandwich and drink combinations are possible. Call on students to suggest combinations. Model listing their suggestions until all possible combinations are represented. Then, with students, count the items in your list to solve the problem. | Distribute manipulative coins and chart paper to each student. Assign each student an amount of money between $0.50 and $1.00. Have students make an organized list of all the possible combinations of coins that equal the given amount. Then ask students to present and explain their organized lists to the group. |

NAME _____ DATE _____

# Lesson 4 Problem-Solving Investigation
## *STRATEGY: Make an Organized List*

**Make an organized list to solve each problem.**

1. **Brianna** has <u>**0.16**</u> of a <u>dollar.</u>
   How many <u>**different**</u> combinations of <u>coins</u> could **she** (Brianna) have?

| Understand | Solve |
|---|---|
| I know:<br><br><br>I need to find:<br><br><br><br> |  |
| **Plan**<br>1 penny = 0.01 of a dollar<br>1 nickel = <u>0.05</u> of a dollar<br>1 dime = <u>0.10</u> of a dollar<br>1 quarter = <u>0.25</u> of a dollar | **Check** |

2. There were <u>**three**</u> races at the track meet.
   The distances were **100** meters long, **800** meters long and **3,200** meters long.
   Suppose **Lucy** ran <u>two</u> of the races.
   What are the possible <u>**total**</u> distances that **she** (Lucy) ran?

| Understand | Solve | | |
|---|---|---|---|
| I know:<br><br><br>I need to find:<br><br><br> | **Lucy's 1st Race** | **Lucy's 2nd Race** | **Total Distance** |
|  | 100m | 800m | 900m |
|  | 100m | 3,200m | 3,300m |
|  | 800m | 3,200m | 4,000m |
| **Plan**<br><br><br><br> | **Check** | | |

# Lesson 5 Convert Metric Units
## English Learner Instructional Strategy

### Vocabulary Support: Build Background Knowledge

Before the lesson, draw a KWL chart on the board. Have students review their My Foldable for Chapter 11, which converted customary units of measurement for weight, length, and capacity. Then say, *Suppose I want to convert feet to inches. Does it take* **more** *or* **fewer** *inches than feet to measure an object?* **more** *When we need more, do we* **multiply** *or* **divide**? **multiply** In the "K" column of the chart, write then say: *To convert larger units of measure to smaller units of measure, multiply.* Have students write other customary conversion examples in "K." Direct students to their Chapter 12 My Foldable and ask, *Do you think* **metric** *units of measurement are converted the same way as* **customary** *units of measurement?* **Yes** Note this in the "W" column. Have students make their Foldable. Complete the "L" column following the lesson.

### English Language Development Leveled Activities

| Emerging Level | Expanding Level | Bridging Level |
|---|---|---|
| **Listen and Identify** | **Building Oral Language** | **Public Speaking Norms** |
| Display a penny and say, *The penny's mass is about one gram.* Have students repeat chorally. Then hold up a book and say, *The book's mass is about 1 kilogram.* Have students repeat chorally. Finally, give each student a penny and a book. Randomly say *gram* or *kilogram.* Then have students chorally repeat the unit of measure you named and hold up the appropriate object. | Collect ten items, some with a mass more than one kilogram and some with a mass that is less. Then draw a T-chart with the headings Grams and Kilograms. Show students an item from the collection, and ask whether they think its mass should be measured in grams or kilograms. Write the item's name in the chart, under the measure they suggest. Then continue with the remaining items. Finally, have students hold each item and say whether its name should move to a different column of the chart. Have students explain their reasoning. | Provide multilingual pairs or groups of students with images of several common household objects that vary in size and mass. (Images can be cut from a magazine or printed from an online source.) Have students estimate the mass of each object. Then have each pair or group present their estimates to the class and explain their reasoning. |

**Teacher Notes:**

NAME _____ DATE _____

# Lesson 5 Guided Writing

## Convert Metric Units

**How do you convert metric units?**

Use the exercises below to help you build on answering the Essential Question. Write the correct word or phrase on the lines provided.

**1.** Rewrite the question in your own words.
_See students' work._
_____

**2.** What key words do you see in the question?
_convert, metric, units_

**3.** Which is a greater unit of measure, one meter or one centimeter?
_1 meter_

**4.** Which is a greater unit of measure, one gram or one kilogram?
_1 kilogram_

**5.** To convert from a larger unit to a smaller unit, use the __multiplication__ operation.

**6.** __Metric__ equivalencies are all multiples of 10, 100, and 1,000.

**7.** 1 kilogram equals 1,000 grams. 2 kilograms equals __2,000__ grams.

**8.** 1 meter equals 100 centimeters. 3 meters equals __300__ centimeters.

**9.** How do you convert metric units?
_Determine which metric equivalency will be used and then use that fact_
_and multiplication to convert the larger unit to the smaller unit._

# Lesson 6 Solve Measurement Problems
## English Learner Instructional Strategy

### Vocabulary Support: Review Basic Vocabulary

Review English terms from the lesson and exercises which may be unfamiliar, such as: *lives, house, poured, lemon juice, pitcher, relay race, runners, bag of potatoes, ribbon, insect, plastic cup, found, sports bag, equipment, golf balls, hockey pucks, necessary.* Display labeled photos or realia in a prominent place in the classroom for students to reference during the lesson to clarify word meaning.

### English Language Development Leveled Activities

| Emerging Level | Expanding Level | Bridging Level |
|---|---|---|
| **Look and Identify** | **Listen and Identify** | **Think-Pair-Share** |
| Review the various units of measurement students have used in this chapter. Show students a metric ruler, a meterstick, containers of various capacities, and items of varying masses. As you point to each unit of measurement, have students identify it either verbally or by pointing to the appropriate written vocabulary word. | Display a variety of measurement tools and containers that students have used throughout the chapter. Use the items to pose measurement problems. For example, say, *If I have a 13-liter bucket and a 9-liter bucket, what is the total amount of liquid the buckets hold?* Have students identify the operation(s) needed to solve the problem. **addition** Then write the problem on the board and solve it with student input. **22 liters** Repeat posing other measurement problems with the items. | Have each student write a real-world word problem that requires converting a larger unit of measure into a smaller unit of measure to solve. Have students choose from the units of measure addressed in this chapter. Then tell students to switch papers with a partner and solve each other's problem. Have partners meet to discuss the problems and check each other's work. |

### Multicultural Teacher Tip

Some students may be used to more formal and stricter learning environments. When groups are formed to do collaborative work, as in Problem Solving Investigations, these students may have a difficult time balancing the social and academic aspects of a cooperative learning activity. A student's response may vary from acting annoyed at any informality, instead wanting to focus only on the task at hand, to being disruptive because he or she views the activity as free social time.

NAME _____ DATE _____

# Lesson 6 Note Taking

## *Solve Measurement Problems*

**Read the question. Write words you need help with and research each word. Use your lesson to write your Cornell notes. Write or draw math examples to explain your thinking. Share your examples with a classmate.**

| | |
|---|---|
| **Building on the Essential Question**<br><br>How do you solve measurement problems in the metric system? | **Notes:**<br><br>Metric units for measurement of length are ___centimeters___ (cm), ___meters___ (m), and ___kilometers___ (km).<br><br>Metric units for measurement of mass are ___grams___ (g) and ___kilograms___ (kg).<br><br>Metric units for measurement of capacity are ___liters___ (L) and ___milliliters___ (mL).<br><br>When solving measurement problems, all values need to be the ___same___ unit ___before___ performing operations.<br><br>There are ___1,000___ milliliters in 1 liter. |
| **Words I need help with:**<br><br>See students' words. | 1 liter − 150 milliliters = ?<br>___1,000___ milliliters − 150 milliliters = ___850___ milliliters<br><br>There are ___10___ millimeters in 1 centimeter.<br><br>5 centimeters + 8 millimeters = ?<br>( ___10___ × 5) centimeters + 8 millimeters = ?<br>___50___ millimeters + 8 millimeters = ___58___ millimeters<br><br>When a problem asks for the **difference** in units, use the operation of ___subtraction___<br><br>When a problem asks for the **total** number of units, use the operation of ___addition___. |
| **My Math Examples:**<br><br>See students' examples. | |

# Chapter 13 Perimeter and Area
## What's the Math in This Chapter?

### Mathematical Practice 5: Use appropriate tools strategically.

Distribute one piece of 1 inch grid paper and 12 1-inch tiles to each student. Say, *Make at 2 different rectangles using your tiles and trace the shape onto the grid paper.* Allow time for students to make rectangles.

Review the definitions of *perimeter* and *area*. Ask, *Why would using grid paper be useful when finding the perimeter or area?* Elicit responses such as, **We can count the number of tile sides around the outside of the shape for the perimeter and the number of squares on the inside for the area.** Allow time for students to find the area and perimeter.

Share results and ask students to identify strategies that they used to help them count. This could include placing X's over each square for the area and using tic marks to count the sides of tiles around the perimeter. Discuss how the grid paper is an appropriate tool/model to use to measure perimeter and area.

Display a chart with Mathematical Practice 5. Restate Mathematical Practice 5 and have students assist in rewriting it as an "I can" statement, for example: **I can use tools/models to understand perimeter and area.** Have students draw or write examples of perimeter and area. Post the examples and new "I can" statement in the classroom.

## Inquiry of the Essential Question:

### Why is it important to measure perimeter and area?

Inquiry Activity Target: **Students come to a conclusion that models can be used to find perimeter and area.**

As an introduction to the chapter, present the Essential Question to students. The inquiry graphic organizer will offer opportunities for students to observe, make inferences, and apply prior knowledge of models representing the Essential Question. As they investigate, encourage students to draw, write, and collaborate with peers to demonstrate their observations and thinking. Then have students present additional questions they may have to a peer to extend discussions.

Regroup students and restate Mathematical Practice 5 and the Essential Question. Pose questions to reflect on what has been learned to guide students in making connections between the Mathematical Practice and the Essential Question.

# Chapter 13 Perimeter and Area

## Inquiry of the Essential Question:

**Why is it important to measure perimeter and area?**

Read the Essential Question. Describe your observations (I see...), inferences (I think...), and prior knowledge (I know...) of each math example. Write additional questions you have below. Then share your ideas and questions with a classmate.

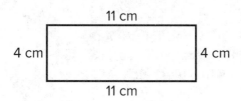

I see ...

I think...

P = Perimeter

P = 11cm + 4cm + 11cm + 4cm = 30cm

I know...

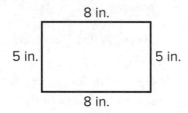

I see ...

I think...

$P = (2 \times \ell) + (2 \times w)$

$P = (2 \times 8) + (2 \times 5)$

P = 16 + 10 or 26 in.

I know...

I see

I think...

A = Area

$A = s \times s$

A = 4 yd × 4 yd

A = 16 sq yd

I know...

Questions I have...

_____

_____

_____

# Lesson 1 Measure Perimeter
## English Learner Instructional Strategy

### Vocabulary Support: Frontload Academic Vocabulary

Before the lesson, write *perimeter* and its Spanish cognate, *perímetro* on the cognate chart. Introduce the term with classroom objects, and provide math examples. Utilize other appropriate translation tools for non-Spanish speaking ELs.

Discuss the multiple meanings of the *terms closed* and *figure*. Create word webs for each term to build background knowledge. Then say, *In this lesson, closed means "without an opening" and figure means "shape."*

Write *rectangular* from Example 1. Circle the suffix *-ular*, and tell students it means "like or related to." Guide students to define *rectangular* as "like or related to a rectangle." Then discuss other words with *-ular*, such as *circular, singular*, and *muscular*.

### English Language Development Leveled Activities

| Emerging Level | Expanding Level | Bridging Level |
|---|---|---|
| **Academic Vocabulary** | **Building Oral Language** | **Show What You Know** |
| Create a 4 by 3 rectangle on a geoboard, and display it for students to see. Trace the outside of the rectangle with your finger and say, *This is the perimeter*. Have students repeat chorally. Then have students join you in counting each unit of the perimeter. On the board, write: perimeter = 14 units. Say, *The perimeter is 14 units*. Have students repeat chorally. Repeat the activity with the perimeter of a square. | Model making a rectangle on a geoboard. Discuss the terms *length* and *width*. Together, count and record the figure's length and width. Model finding the perimeter using the formula for finding the perimeter of figures. In pairs, have students create a rectangle on a geoboard with a rubber band. Ask them to find its perimeter. Allow pairs to present their rectangle using these sentence frames: **The length is ____. The width is ____. The perimeter is ____.** | Assign each student a perimeter length in inches or centimeters units. Then have students use a geoboard or paper, pencil, and a ruler to create a shape that has the assigned perimeter. Ask students to present their shape to the class. Have them tell why they selected their shape and how they knew what length to use for each side. |

### Teacher Notes:

NAME _____ DATE _____

# Lesson 1 Vocabulary Definition Map

*Measure Perimeter*

Use the definition map to write a description and list characteristics about the vocabulary word or phrase. Write or draw math examples. Share your examples with a classmate.

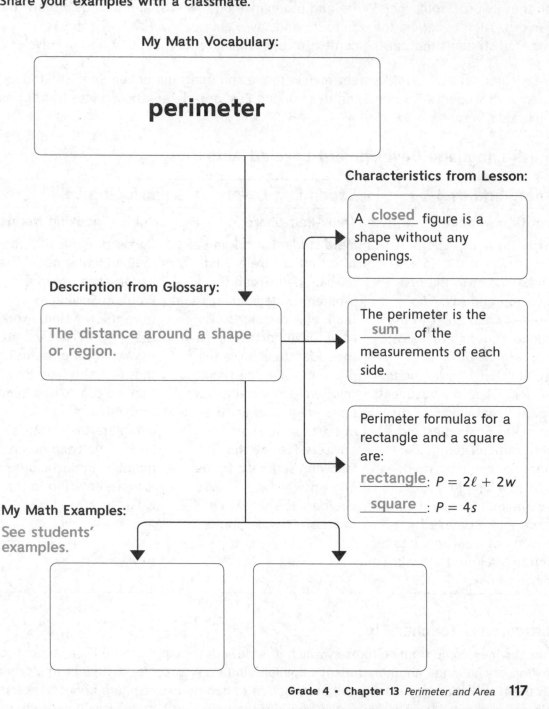

My Math Vocabulary:

## perimeter

Characteristics from Lesson:

A __closed__ figure is a shape without any openings.

Description from Glossary:

The distance around a shape or region.

The perimeter is the __sum__ of the measurements of each side.

Perimeter formulas for a rectangle and a square are:

__rectangle__: $P = 2\ell + 2w$

__square__ : $P = 4s$

My Math Examples:

See students' examples.

# Lesson 2 Problem-Solving Investigation
## Strategy: Solve a Simpler Problem
### English Learner Instructional Strategy

## Collaborative Support: Native Language Peers/Aides

Before the lesson, discuss these terms from the Learn the Strategy example: *route, block*. First, write *route*. Tell students that, depending on dialect, *route* can be a homograph for both these words: *rout, root*. Write and pronounce both words, and provide visuals to clarify meanings. Then underline the *e* in rout*e* and say, *Route with an "e" usually has to do with travel*. Tell students that, in the context of this lesson, route means "a line of travel."

Have a bilingual aide or older peer review the lesson using the online Spanish Student Edition with students. For non-Spanish speaking ELs, provide an appropriate translation tool for the aide to utilize.

## English Language Development Leveled Activities

| Emerging Level | Expanding Level | Bridging Level |
|---|---|---|
| **Word Knowledge** | **Think-Pair-Share** | **Public Speaking Norms** |
| Write these equations: $4 \times 5 = 20$; $3(5 \times 1) \times (6 + 2) \div 4 = 30$. Point to the first equation and say, *This is simpler*. Then point to the second equation and say, *This is more difficult*. Explain that *simpler* means "easier." Then show students pictures that illustrate other examples of simpler and more difficult things. For example, compare items such as a simple maze and a complex maze, a stick figure and a detailed drawing of a person, and a picture book and a thick chapter book. | Have students work in pairs to solve Exercise 4. First, have students read the problems independently and think of a plan for solving. Next, have partners discuss their ideas and solve the problem using the simplest solution idea. Finally, have pairs meet with another set of partners to share answers. Provide the following sentence frames: **The answer is _____. We decided to solve it by _____ because it was simpler.** | Divide students into three multilingual groups. Then assign each group a Problem Solving exercise from the My Homework of the lesson. Ask each group to prepare a presentation that explains how they applied the solve a simpler problem strategy to complete their exercise. Encourage each group member to participate in the presentation to the class. |

## Multicultural Teacher Tip

Some ELs may come from cultures in which it is viewed as inappropriate for a student to question or challenge another student's opinion, and even worse for a student to disagree with the teacher. These students will be reluctant to debate or speculate about answers during class discussions. They may also consider their input to be unworthy unless what they add to the discussion is a knowledgeable, well-prepared, and precisely correct answer.

NAME _____ DATE _____

# Lesson 2 Problem-Solving Investigation
## STRATEGY: Solve a Simpler Problem

Solve each problem by solving a simpler problem.

1. Clarissa has **four** **pictures** that are each the **size** of the one **shown**.
   What will be the <u>perimeter</u> of the <u>rectangle</u> formed if the four pictures are
   laid end to end as show?

   5 in.     5 in.     5 in.     5 in.

   3 in.                                    3 in.

   | Understand | Solve |
   |------------|-------|
   |            |       |
   | **Plan**   | **Check** |
   |            |       |

2. Mr. and Mrs. Lopez are putting <u>square</u> tiles on the floor in their bathroom.
   **They** (Mr. and Mrs. Lopez) can fit **6 rows** of **4 tiles** in the bathroom.
   How many **tiles** do they **need** to buy?
   If **each** tile costs **$5**, what is the **total cost**?

   | Understand | Solve |
   |------------|-------|
   |            |       |
   | **Plan**   | **Check** |
   |            |       |

# Lesson 3 Inquiry/Hands On: Model Area
## English Learner Instructional Strategy

### Sensory Support: Realia

Before the lesson, write *area* and its Spanish cognate, *área,* on a cognate chart. Have students write the term *area* and its translation in their native language on an index card. Then explain the term and provide a math example for students to include on their cards. Have students store their cards in their math journals for future review.

Then write: *unit square, square unit.* Help students differentiate these terms. Explain that in *unit square,* the words *unit* and *square* work together to name a type of figure. Show students a sheet of grid paper. Point to the squares on the paper and say, *Unit square is the name for each of these figures.* Show students that one unit forms each side of each figure. Then explain that in *square unit, square* is used as an adjective, or a describing word. Say, *Square describes the shape of a unit that is used to measure area.* Help students remember that a square unit measures area by pointing out that a square also *has* area.

### English Language Development Leveled Activities

| Emerging Level | Expanding Level | Bridging Level |
|---|---|---|
| **Non-Transferrable Sounds** | **Exploring Language Structure** | **Turn and Talk** |
| Write *cover* and draw a line between the *o* and *v* to divide the word into syllables. Point to the first syllable and say: /kə/. Demonstrate that the *c* says /k/ not /s/ and the *o* says *uh* not /ō/ or /o/. Say /kə/ again and have students chorally repeat. Then point to the second syllable and say: /ver/. Demonstrate that *v* says /v/ not /b/, /w/, or /f/. (Because /v/ is not used in all languages, students may need extra practice with saying this sound.) Say /ver/ again and have students chorally repeat. Then say: /kə ver/. Have students chorally repeat. | Have students do the Emerging Level activity. Then discuss multiple meanings for *cover.* Point out that most meanings relate in some way to the act of laying one thing over the top of another. Demonstrate by saying, for example, *I will cover the pencil,* and laying a piece of paper over it. Show a picture of a bed that has a cover, or blanket, on top of it. Invite students to tell the color of their bed covers using this sentence frame: **My cover is ____.** Then point out how unit squares cover, or blanket the rectangles pictured with the Draw It activities. | Help students recall what they know about *factors.* Write 12 and ask, *What are the factors of 12?* 1 × 12, 2 × 6, 3 × 4 Then read aloud Write About It Exercise 13 with students. Have students turn to a neighbor and share their ideas about the answer to the question. After they have had time to discuss, have students work individually to write their responses. Provide this sentence starter for them to use: **The two rectangles [definitely/may not] have the same length and width because ____.** Invite students to share their responses with the class. |

### Teacher Notes:

NAME _____ DATE _____

# Lesson 3 Guided Writing

## Inquiry/Hands On: Model Area

**How do you model area?**

**Use the exercises below to help you build on answering the Essential Question. Write the correct word or phrase on the lines provided.**

1. Rewrite the question in your own words.
   See students' work.

2. What key words do you see in the question?
   area, model

3. A square is a rectangle that has _four_ sides of the same length. The length and width are _equal_ in a square.

4. The side length of the square below is _1_ unit(s).

5. The side length of the square below is _2_ unit(s).

6. _Area_ is the number of square units needed to cover the inside of a region or plane figure without any overlap.

7. A unit square is different from a square unit. A _unit square_ is a square with a side length of one unit. _Square unit_ is a unit for measuring area.

8. The area of the square below is _1_ square unit(s).

9. The area of the square below is _4_ square unit(s).

10. How do you model area?
    Identify the length and width of the rectangle. Draw unit squares to model the rectangle. Find the total number of square units that make up the rectangle to find the area.

Grade 4 • Chapter 13 *Perimeter and Area*  **119**

# Lesson 4 Measure Area
## English Learner Instructional Strategy

### Language Structure Support: Communication Guide

After students complete Exercises 7–10 on their own, have them explain their responses to a partner. Direct partners to take turns explaining, using this communication guide:

**The unknown measures are \_\_\_\_, \_\_\_\_, and \_\_\_\_.**
**I know this because \_\_\_\_ divided by \_\_\_\_ is \_\_\_\_.**

Tell partners to make sure they agree on the response to each problem before moving on to the next one.

### English Language Development Leveled Activities

| Emerging Level | Expanding Level | Bridging Level |
|---|---|---|
| **Word Knowledge** | **Building Oral Language** | **Synthesis** |
| Demonstrate the meaning of *area* by walking or pointing to different areas of the classroom and telling how they are used. For example, say, *We hang coats in this area. We keep books in this area. We keep art supplies in this area.* Then explain that area is also used to describe the measurement of space inside a shape. Draw a 5-inch by 5-inch square, and divide it into 25 square inches. With students, count the square inches aloud. Then say, *The area is 25 square inches.* Have students repeat chorally. Repeat the exercise with other squares and rectangles. | Draw a rectangle. Label its length 8 feet and its width 3 feet. Then write: $A = \ell \times w$. Say, *Area equals length times width.* Have students repeat chorally. Model using this formula to find the area of your rectangle: 8 ft × 3 ft = 24 sq ft. Then draw a square. Label one side 6 cm. Then write: $A = s \times s$. Say, *Area equals side times side.* Have students repeat chorally. Model using this formula to find the area of your square: 6 cm × 6 cm = 36 sq cm. Finally, assign each student measurements for a rectangle or square and have them find the area. | Have students estimate the area of a book cover, computer screen, or desktop to the nearest inch or centimeter. Then have students use a customary or metric ruler to find accurate measures and calculate the area. Finally, have students use graph paper to draw a model of the square or rectangle they measured. Tell students to label the model appropriately and to write the formula they used for finding the area of the shape. |

**Teacher Notes:**

NAME _____ DATE _____

# Lesson 4 Multiple Meaning Word
## *Measure Area*

Complete the four-square chart to review the multiple meaning word or phrase.

| Everyday Use | Math Use in a Sentence |
|---|---|
| Sample answer: A part of a city, or state, or the county is an area of that land. For example, the rural areas of the state. | Sample sentence: The area is the amount of space a figure takes up. |

**area**

| Math Use | Example From This Lesson |
|---|---|
| The number of square units needed to cover the inside of a region or plane figure without any overlap. | Sample answer: The area of the rectangle below is 6 square units. |

**Write the correct term on each line to complete the sentence.**

To find the  area  of a figure you can count the number of unit squares or multiple the  length  by  width .

# Lesson 5 Relate Area and Perimeter
## English Learner Instructional Strategy

### Vocabulary Support: Making Connections

Before the Developing Vocabulary, discuss what it means to "relate area and perimeter." On the board, write: Relate _____ to _____. Then say, *When we **relate** two things, we tell how they are connected or similar.* As an example, use the words soccer and football in the sentence frame. Have students help you list ways soccer and football are similar. For example, both are sports, both are played with a ball, both involve kicking, and so on. Next, have students suggest differences between soccer and football, for example shape of the balls, style of uniforms, etc. Then ask, *How are **area** and **perimeter** related?* Display a Venn diagram and continue with the Vocabulary Review Activity.

### English Language Development Leveled Activities

| Emerging Level | Expanding Level | Bridging Level |
|---|---|---|
| **Listen and Identify**<br><br>Draw a rectangle and label it 5 cm by 7 cm. Trace the figure with your finger and say, *Perimeter is the distance around.* Gesture toward the interior of the rectangle and say, *Area is the space inside.* Write: 7 cm + 5 cm + 7 cm + 5 cm = 24 cm; 7 cm × 5 cm = 35 cm. Point to the first equation and ask, *Is this the perimeter or the area?* **perimeter** Point to the second equation and ask, *Is this the perimeter or the area?* **area** Repeat the activity with rectangles of different sizes and have students find the area and perimeter for each. | **Report Back**<br><br>Divide students into pairs and give each pair two rectangles with length and width labeled. (Be sure that each pair of rectangles has either the same perimeter or same area). Each partner finds the area and perimeter of one rectangle, then partners compare results. Have students use the following sentence frames to report their findings: **My rectangle's perimeter is _____ units. My rectangle's area is _____ square units. The rectangles have the same _____ (perimeter/area).** | **Show What You Know**<br><br>Give each student a copy of the *Centimeter Grid Paper* manipulative master found online in Program Resources, and have them draw a rectangle. Then tell students to switch papers with a partner. Have students find the perimeter in units and area in square units of the rectangle they received. Then have partners check each other's work. |

### Teacher Notes:

NAME _____ DATE _____

# Lesson 5 Note Taking

## *Relate Area and Perimeter*

Read the question. Write words you need help with and research each word. Use your lesson to write your Cornell notes. Write or draw math examples to explain your thinking. Share your examples with a classmate.

| Building on the Essential Question | Notes: |
|---|---|
| How are area and perimeter related? | Area is the number of <u>square units</u> needed to cover the <u>inside</u> of a region or plane figure without any <u>overlap</u>. <br><br> Perimeter is the <u>distance around</u> a shape or region. <br><br> Two figures <u>can</u> have the same <u>perimeter</u> and different areas. <br><br>  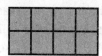 |

**Words I need help with:**

See students' words.

Perimeter: <u>12</u> units    Perimeter: <u>12</u> units

Area: <u>9</u> square units    Area: <u>8</u> square units

Two figures <u>can</u> have the same <u>area</u> and different perimeters.

Perimeter: <u>14</u> units    Perimeter: <u>16</u> units

Area: <u>12</u> square units    Area: <u>12</u> square units

**My Math Examples:**

See students' examples.

Grade 4 · **Chapter 13** *Perimeter and Area*  **121**

# Chapter 14 Geometry

## What's the Math in This Chapter?

### Mathematical Practice 7: Look for and make use of structure.

Draw or display pictures of acute and obtuse triangles and acute and obtuse angles on large sticky notes. Display randomly on the board. Ask, *How could I sort these?* Allow students time to think and to share their ideas with each other. Then sort the angles and triangles as recommend by the students. The goal of the sorting is for students to classify the figures by their geometric structure, sorting all of the acute items together and all of the obtuse items together.

Say, *We used the* **structure** *of these geometric figures to help us sort them. We can classify the two groups as acute or obtuse based on their angle measurement.* Label each group accordingly. Discuss that just like Mathematical Practice 7, they looked for and made use of the structure of the geometric figures to sort them.

Display a chart with Mathematical Practice 7. Restate Mathematical Practice 7 and have students assist in rewriting it as an "I can" statement, for example: **I can see how shapes are alike and different.** Post the new "I can" statement in the classroom.

## Inquiry of the Essential Question:

### How are different ideas about geometry connected?

Inquiry Activity Target: **Students come to a conclusion that geometric shapes have similarities.**

As an introduction to the chapter, present the Essential Question to students. The inquiry graphic organizer will offer opportunities for students to observe, make inferences, and apply prior knowledge of shapes representing the Essential Question. As they investigate, encourage students to draw, write, and collaborate with peers to demonstrate their observations and thinking. Then have students present additional questions they may have to a peer to extend discussions.

Regroup students and restate Mathematical Practice 7 and the Essential Question. Pose questions to reflect on what has been learned to guide students in making connections between the Mathematical Practice and the Essential Question.

NAME _____ DATE _____

# Chapter 14 Geometry

*Inquiry of the Essential Question:*

**How are different ideas about geometry connected?**

Read the Essential Question. Describe your observations (I see...), inferences (I think...), and prior knowledge (I know...) of each math example. Write additional questions you have below. Then share your ideas and questions with a classmate.

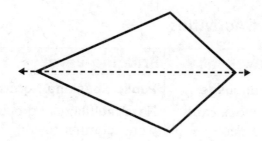

I see ...

I think...

I know...

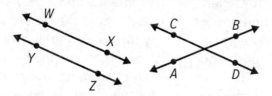

I see ...

I think...

I know...

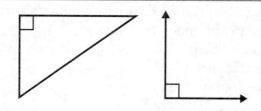

I see ...

I think...

I know...

Questions I have...

_____

_____

_____

# Lesson 1 Draw Points, Lines, and Rays
## English Learner Instructional Strategy

### Vocabulary Support: Frontload Academic Vocabulary

Before the lesson, write *line* and *line segment* and their Spanish cognates, *línea* and *segmento de línea*, respectively on a cognate chart. Introduce each word, and provide math examples. Utilize other appropriate translation tools for non-Spanish speaking ELs. Discuss multiple meanings for the word *line*, especially those associated with subjects students learn about in school, such as letter writing, drama, poetry, and sports. Also, tell students that *segment* and *part* are synonyms.

### English Language Development Leveled Activities

| Emerging Level | Expanding Level | Bridging Level |
|---|---|---|
| **Word Knowledge** | **Building Oral Language** | **Public Speaking Norms** |
| Draw an example of an *endpoint, line, line segment,* and *ray*. Point to each example, and say the figure's name. Have students repeat chorally. Then point to the ray's endpoint and say, *This is the* **endpoint**. Have students repeat chorally. After you have named all the figures, randomly point to them and have students identify the figures chorally. Once students demonstrate understanding, ask volunteers to come to the board and write the correct name beneath each figure. | Gather enough index cards to distribute one per student. On each card, draw an endpoint, a line, a line segment, or a ray. Then review the definitions for *point, line, line segment,* and *ray*. For English and Spanish definitions, have students refer to the Glossary located at the back of their student book. Finally, distribute the cards to students. Have them take turns holding up a card and identifying and describing the figure. Have students use this sentence frame: **I know this is a ____ because it has ____ .** | Have multilingual groups work together to compare and contrast the lesson vocabulary: *lines, line segments,* and *rays*. Ask students to create a chart with supporting illustrations that organizes their findings. Direct students to use the following terms: *points, extends, opposite, between, endpoints*. Finally, ask each group to present its compare/contrast chart to the class. |

**Teacher Notes:**

NAME _____ DATE _____

# Lesson 1 Vocabulary Chart

## *Draw Points, Lines, and Rays*

Use the three-column chart to organize the vocabulary in this lesson.
Write the word in Spanish. Then write the correct terms to complete
each definition.

| English | Spanish | Definition |
|---------|---------|------------|
| **line** | recta | A <u>straight</u> set of points that extend in <u>opposite</u> directions without ending. |
| **line segment** | segmento de recta | A part of a <u>line</u> between two <u>endpoints</u>. The length of the line segment can be <u>measured</u>. |
| **endpoint** | extremo | The <u>point</u> at either end of a <u>line segment</u> or the <u>point</u> at the beginning of a <u>ray</u>. |
| **point** | punto | An <u>exact</u> location in space that is represented by a <u>dot</u>. |
| **ray** | semirrecta | A part of a <u>line</u> that has one <u>endpoint</u> and extends in one direction without ending. |

# Lesson 2 Draw Parallel and Perpendicular Lines

## English Learner Instructional Strategy

### Sensory Support: Mnemonic/Illustrations

Before the lesson, write *parallel* and *perpendicular* and their Spanish cognates, *paralelo* and *perpendicular*, respectively on the cognate chart. Model the meaning of each word, and provide math examples. Then point out that the three "l"s in *parallel* are a clue to the word's definition—they actually form parallel lines and also resemble the symbol for *parallel* (‖). Next, explain the terms: *cross, intersect(s)/intersecting*. Discuss multiple meanings for *cross* using pictures. Then draw a pair of crossing lines. Use them to show students that when lines cross each other, they meet at a point. Finally, tell students that *cross* and *intersect* are synonyms.

### English Language Development Leveled Activities

| Emerging Level | Expanding Level | Bridging Level |
|---|---|---|
| **Act It Out**<br><br>Each student will need two pencils for this exercise. Hold two pencils parallel to one another. Say, *The pencils are parallel*. Have students repeat chorally while mimicking your action. Then cross the pencils so they create a square angle. Say, *The pencils are perpendicular*. Have students repeat chorally while mimicking your action. Finally, hold the pencils so that they are crossed yet not perpendicular. Say, The *pencils are intersecting*. Have students repeat chorally while mimicking your action. Then randomly say each term and have students model it. | **Building Oral Language**<br><br>Use two pencils or rulers to demonstrate *parallel, perpendicular,* and *intersecting lines*. Say each term as you demonstrate it, and have students repeat chorally. Then have students identify examples of parallel, perpendicular, and intersecting lines in the classroom. For example, two sides of a doorway are parallel; one side of the doorway and the floor are perpendicular; and many lines on a map intersect. | **Making Connections**<br><br>Have each student create a drawing that incorporates *parallel, perpendicular,* and *intersecting* line segments. Then have students show their drawings to the class. The presenter will ask their classmates to identify examples of parallel, perpendicular, and intersecting line segments in their drawing. |

## Teacher Notes:

NAME _____ DATE _____

# Lesson 2 Vocabulary Definition Map
## *Draw Parallel and Perpendicular Lines*

Use the definition map to write a description and list characteristics
about the vocabulary word or phrase. Write or draw math examples.
Share your examples with a classmate.

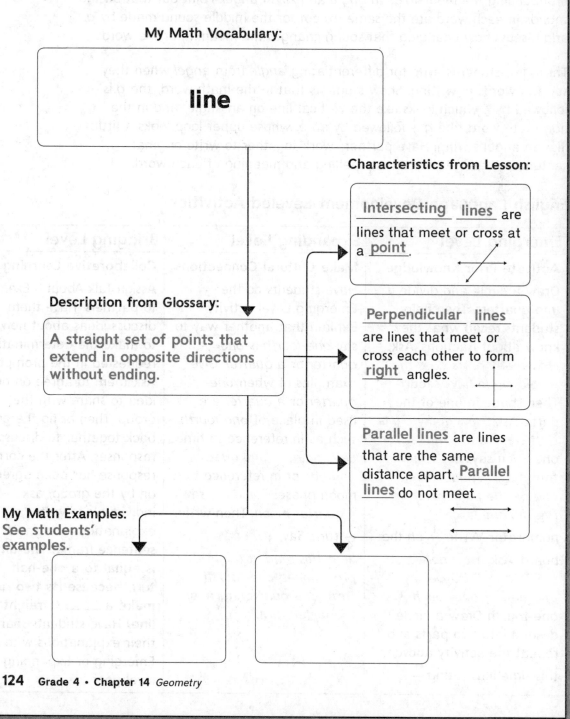

**My Math Vocabulary:**

line

**Characteristics from Lesson:**

_Intersecting_ _lines_ are
lines that meet or cross at
a _point_ .

_Perpendicular_ _lines_
are lines that meet or
cross each other to form
_right_ angles.

_Parallel_ _lines_ are lines
that are the same
distance apart. _Parallel_
_lines_ do not meet.

**Description from Glossary:**

A straight set of points that
extend in opposite directions
without ending.

**My Math Examples:**
See students'
examples.

# Lesson 3 Inquiry/Hands On:
# Model Angles
## English Learner Instructional Strategy

### Sensory Support: Pictures/Mnemonic

Before the lesson, label and display pictures of an angle and an angel. Point to the label on each picture and read it aloud. Take special care in pronouncing the medial /g/ in *angle* and /j/ in *angel*. Point out that all the sounds in each word are the same except for the middle sound made by *g*, and discuss how changing that sound changes the meaning of the word.

Then give students a tip for differentiating *angle* from *angel* when they see the words in writing. Show students that in the math word, the *g* is followed by *l*, which looks like the vertical line on an angle, and in the non-math word, the *g* is followed by an *e*, whose upper loop looks a little like an angel's wing. Have partners work together to write original sentences that use the correct spelling and meaning of each word.

### English Language Development Leveled Activities

| Emerging Level | Expanding Level | Bridging Level |
|---|---|---|
| **Activate Prior Knowledge** | **Make Cultural Connections** | **Collaborative Learning** |
| Draw a circle and divide it into quarters. Then help students recall what they know about fractions. Ask, *How many parts does the whole circle have?* **four** Then shade in one of the parts. Ask, *How many parts of the circle are shaded?* **one** Then say, *Let's show this in a fraction. Should one be the numerator or the denominator?* **numerator** Write $\frac{1}{4}$ on the board. Ask, *How do we say this fraction? Do we say one-half or one-fourth?* **one-fourth** Draw a circle divided into two parts and repeat the activity above, this time illustrating $\frac{1}{2}$. | Have students do the Emerging Level activity. Explain that another way to say *one-fourth* is *one-quarter* or *a quarter*. Give examples of when *one-quarter* or *a quarter* are used in place of *one-fourth*, such as in reference to time: *Let's meet in one quarter of an hour.* or in reference to a moon phase: *Tonight I saw a quarter moon.* Then write: $\frac{1}{4}$ *turn.* Say, *When you hear this read, it might be pronounced* **one-fourth turn, one-quarter turn,** or *a quarter turn.* | Assign Talk About It Exercise 1 to partners. Have them discuss ideas about how to make the determination requested in the prompt. Ask them to agree on one idea to share with the group. Then bring the group back together to discuss responses. After the correct response has been agreed on by the group, ask individuals to write explanations using this sentence frame: **The angle is equal to a one-half turn because its two rays make a ____. (straight line)** Have students share their explanations with an Emerging or Expanding level peer. |

### Teacher Notes:

NAME _____ DATE _____

# Lesson 3 Guided Writing

## Inquiry/Hands On: Model Angles

**How do you model angles?**

Use the exercises below to help you build on answering the Essential Question. Write the correct word or phrase on the lines provided.

1. Rewrite the question in your own words.
   See students' work.

   _____

2. What key words do you see in the question?
   angles, model

3. Identify if each situation below models a $\frac{1}{4}$ turn, a $\frac{1}{2}$ turn, or a full turn.

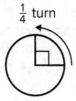   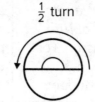   

$\frac{1}{4}$ turn          $\frac{1}{2}$ turn          1 full turn

   a. Extend both arms together out in front of you and then move one arm straight out to the side. $\frac{1}{4}$ turn

   b. A clock's minute hand was on the 12, then moved to the 6. $\frac{1}{2}$ turn

4. A  ray  is a part of a line that has one endpoint and extends in one direction without ending.

5. An  angle  is a figure that is formed by two rays with the same endpoint.

6. Identify if each angle models as a $\frac{1}{4}$ turn, a $\frac{1}{2}$ turn, or a full turn between the two rays.

   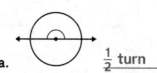          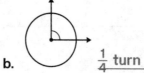

   a.          $\frac{1}{2}$ turn          b.          $\frac{1}{4}$ turn

7. How do you model angles?
   Draw one ray horizontal. Draw a second ray, with the same
   endpoint, that creates the turn of the angle you are modeling.

# Lesson 4 Classify Angles

## English Learner Instructional Strategy

### Vocabulary Support: Multiple Meaning Words

Before the lesson, write *angle, acute angle,* and obtuse angle and their Spanish cognates, *ángulo, ángulo agudo,* and *ángulo obtuso,* respectively on a cognate chart. Introduce each word, and provide math examples. Discuss multiple meanings for *angle, acute,* and *obtuse.* Finally, have students preview pictures of angles in this lesson's Student Edition. Point out the red arc inside the acute and obtuse angles, and use pencils to model that the red arc shows how far the angle's rays have turned from each other.

### English Language Development Leveled Activities

| Emerging Level | Expanding Level | Bridging Level |
|---|---|---|
| **Listen and Write** | **Building Oral Language** | **Synthesis** |
| On the board, draw a right angle. Point to the angle and say, *This is a right angle.* Have students repeat chorally. Next, use the same procedure to model and identify an *acute angle* and an *obtuse angle.* Finally, have three or four students come to the board. Ask them to draw a right angle, an acute angle, or an obtuse angle. If students need guidance, point to your example drawing and repeat the angle's name. Continue this activity until all students have participated. | Review the definitions and examples for *right angle, acute angle,* and *obtuse angle* from the lesson. For English and Spanish definitions, have students refer to the Glossary in their Student Edition. Then use the hands of a demonstration clock to form different types of angles. Have students identify each one as a right angle, an acute angle, or an obtuse angle. Ask students to explain their responses using this sentence frame: **The angle is ___ because ___.** | Divide students into pairs, and distribute one manipulative clock to each pair. Have students use the clock to find three times: one at which the clock's hands form *a right angle,* one at which the hands form an *acute angle,* and one at which the hands form an *obtuse angle.* Then ask each pair to present their times and corresponding angles to the class. |

**Teacher Notes:**

NAME _____ DATE _____

# Lesson 4 Concept Web
## *Classify Angles*

Use the concept web to write if the angle is classified as *acute, obtuse,* or *right* based on the drawing or measurement given.

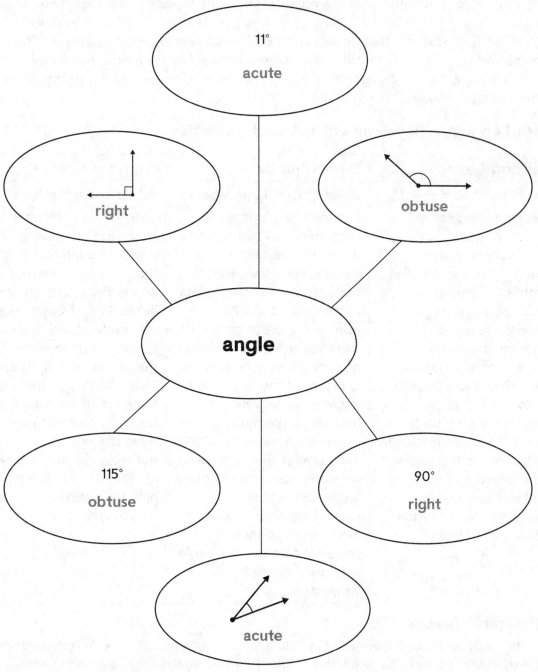

# Lesson 5 Measure Angles

## English Learner Instructional Strategy

### Vocabulary Support: Build Background Knowledge

Before the lesson, discuss these terms: *line up, straightedge, align(s)*.

First, demonstrate multiple meanings for *line up*. Say, *I will **line up** these objects.* Arrange the objects in a line. Then, invite a few students to the front of the classroom. Arrange them in a line, and then say, ***Line up**, students! Please stand in a line.* Finally, model lining up a ruler, yardstick, or protractor with the edges of different objects in the classroom, saying, for example, *I will **line up** the ruler with the desk's edge. I will **align** the yardstick with the chalkboard's edge.* Explain that, in this context, *line up* and *align* are synonyms. Then write: *straightedge.* Show students that this is a compound word by drawing a line between *straight* and *edge.* Say, *A **straightedge** is an **edge** used to make a **straight** line.* Show examples of straightedges.

### English Language Development Leveled Activities

| Emerging Level | Expanding Level | Bridging Level |
|---|---|---|
| **Show What You Know** | **Building Oral Language** | **Public Speaking Norms** |
| Display a protractor and say, *This is a **protractor**.* Have students repeat chorally. Then explain that a protractor is used for measuring angles. Demonstrate how the protractor is used by drawing an angle on the board. Then show students how to: 1. Line up the protractor. 2. Line up the angle. 3. Measure the angle. Finally, draw other angles on the board and have volunteers use a protractor to measure them. Provide guidance as needed. | Distribute a protractor to each student. Point out these features: the center; the zero on either end; the two sets of tic marks. Write: *acute = less than 90°; obtuse = greater than 90°.* Next, draw an acute angle, and review how to measure it. Explain how, because the angle is acute, you know that the angle must measure between 1 and 89°. Then repeat the exercise with other acute and obtuse angles. Have students identify each angle and its measure using these sentence frames: **The angle is [acute/obtuse] _____ . Its measure is _____ .** | Distribute a protractor to each student. Have each student identify an angle in the classroom, such as the angle at which a chair leg bends away from its seat or the angle at which a stapler bends. Ask students to find the measure of their angle. Then encourage them to think creatively about the best way to measure it. Have students present to the class the angle they identified, its measure, and how they performed the measurement. |

### Multicultural Teacher Tip

During the lesson, you may experience ELs who appear to listen closely to your instructions and exhibit verbal and/or nonverbal confirmation that they understand the concepts. However, it becomes clear during the Guided or Independent Practice parts of the lesson that the students did not actually understand. This may be due to a student coming from a culture in which the teacher is regarded as a strong, perhaps even intimidating, authority figure. They may be reluctant to ask questions, considering it impolite to do so and an implication that the teacher is failing.

NAME _____ DATE _____

# Lesson 5 Multiple Meaning Word
## *Measure Angles*

Complete the four-square chart to review the multiple meaning word.

**Everyday Use**

Sample answer: A certain way of approaching an issue or problem. You can approach a problem from a different angle.

**Math Use in a Sentence**

Sample sentence: An angle is the shape formed when two straight lines meet.

**angle**

**Math Use**

A figure that is formed by two rays with the same endpoint.

**Example From This Lesson**

Sample answer: A triangle has three angles. A rectangle has four angles.

**Write the correct term on each line to complete the sentence.**

Three types of angles are <u>obtuse</u>, <u>acute</u>, and <u>right</u>.

# Lesson 6 Draw Angles

## *English Learner Instructional Strategy*

### Language Structure Support: Echo Read

Before the lesson, write *vertex* and its Spanish cognate *vértice* on a cognate chart. Introduce the term, and provide a math example. Then write the plural form *vertices*. Pronounce *vertices,* and tell students it means "more than one vertex." Draw a math example.

During the lesson, read aloud Exercise 12 and have students echo read. Ensure they understand that the exercise requires them to draw two angles and explain why their second angle cannot be Julia's. Provide this sentence frame to help students form their explanations: **My angle cannot be Julia's angle because my angle measures _____ .**

### English Language Development Leveled Activities

| Emerging Level | Expanding Level | Bridging Level |
|---|---|---|
| **Listen, Look, and Draw** | **Building Oral Language** | **Synthesis** |
| Draw an angle on the board. Point to each ray and say, *This is a ray.* Have students repeat chorally. Then point to the vertex and say, *This is the vertex.* Have students repeat chorally. Next, show how to draw a 30° angle. Draw one ray of the angle, and mark the endpoint. Place the protractor along the ray as you would to measure an angle. On the protractor, find 30°. Then mark the spot, and use a ruler to draw the ray that connects the endpoint with the mark. Have students mimic each step as you perform it, drawing the angle on their own paper. | Write then say aloud these steps as you model drawing an angle: *1. Draw one ray. 2. Measure the angle. 3. Draw the other ray.* Then have three students at a time come to the board. Name an angle measure, and have each student perform the steps for drawing the angle. Tell students to describe each step as they perform it. | Write and read aloud this formula and example: *Your age + the number of letters in your name + the day of the month on which your birthday falls. For example, 10 years + 7 letters + 20th day = 37 degrees.* Have students use the formula to come up with a number they will use as an angle measure. Then have them draw the angle. Finally, ask students to make up their own formulas for determining an angle measure. Then have them exchange formulas with a partner. Tell students to draw their angles and check each other's work. |

**Teacher Notes:**

NAME _____ DATE _____

# Lesson 6 Vocabulary Cognates
## *Draw Angles*

Use the Glossary to define the math word in English and in Spanish in the word boxes. Write a sentence using your math word.

| **angle** | **ángulo** |
|---|---|
| **Definition**<br><br>A figure that is formed by two rays with the same endpoint. | **Definición**<br><br>Figura formada por dos semirrectas con el mismo extreme. |
| **My math word sentence:**<br>Sample answer: An angle can be measured using a protractor. ||

| **ray** | **semirrecta** |
|---|---|
| **Definition**<br><br>A part of a line that has one endpoint and extends in one direction without ending. | **Definición**<br><br>Parte de una recta que tiene un extreme y se extiende sin fin en una dirección. |
| **My math word sentence:**<br>Sample answer: To draw an angle, you need to first draw a ray. ||

# Lesson 7 Solve Problems with Angles

## English Learner Instructional Strategy

### Vocabulary Support: Activate Prior Knowledge

Preview the following lesson terms: *decomposed* and *non-overlapping*. Assess students prior knowledge of *decompose*, having them recall examples of how they have decomposed numbers in past lessons. model the definition for those students unfamiliar with the term. Then write *non-overlapping* on the board. Circle *non-*, and tell students this is a prefix meaning "not." Then underline *overlap*, and demonstrate its meaning. Show students examples of items in the classroom that overlap each other, such as papers on your desk or pictures posted in a display. Say, for example, *These pictures overlap*, showing how one picture lies over or partly covers the other. Then point to *non-overlapping* on the board and say, *If objects are non-overlapping, they do **not** overlap*. Encourage students to point out examples in the classroom.

### English Language Development Leveled Activities

| Emerging Level | Expanding Level | Bridging Level |
|---|---|---|
| **Academic Vocabulary** | **Number Sense** | **Number Game** |
| Display a set of 3 connecting cubes, and write the number 3 on the board. Then display a set of 5 connecting cubes, and write the number 5 on the board. Finally, connect the first and second sets of cubes and say, *The cubes are combined*. Have students repeat chorally. On the board, write 3 + 5 = 8, and say, *The **combined** number of cubes is 8.* Explain that combined means "total." Then draw two angles, 70° and 10°, which represent a combined angle of 80°. Discuss how the two angles are combined to measure 80°. | On the board, draw adjacent angles that measure 45° and 65°. Demonstrate how to find the combined angle measure by adding 45° and 65°. Then say, *The **combined** angle measure is 110°.* Next, draw adjacent angles with other measures, and have students identify the combined angle measure using this sentence frame: **The combined angle measure is _____.** Finally, provide a combined angle measure of 125° and this measure of one of the angles: 75°. Have students subtract to find the unknown angle measure. **50°** | Draw an angle on the Draw an angle on the board, and divide it into two adjacent angles. Label the angle's total measure and the measure of one of the smaller angles. Then have students race to identify the measure of the unknown angle. The first student to correctly identify the unknown angle measure gets to draw and label the angle for the next round. Repeat until all students have had the opportunity to draw and label an angle. |

**Teacher Notes:**

NAME _____ DATE _____

# Lesson 7 Note Taking
## *Solve Problems with Angles*

Read the question. Write words you need help with and research each word. Use your lesson to write your Cornell notes. Write or draw math examples to explain your thinking. Share your examples with a classmate.

| | |
|---|---|
| **Building on the Essential Question**<br><br>How can you solve problems with angles? | **Notes:**<br><br>To decompose a number means to <u>break</u> a number into <u>different</u> parts.<br><br>$\underline{90} = 50 + 40$<br><br>An angle can also be decomposed into <u>different</u> parts.<br><br>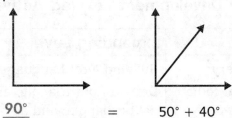<br><br>$90°$      $=$      $50° + 40°$ |
| **Words I need help with:**<br>See students' words. | A variable is a letter or symbol used to represent an <u>unknown</u> quantity.<br><br>When solving a problem to find an angle measure, you can use a <u>variable</u> to represent the unknown angle measure.<br><br>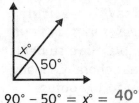<br><br>$90° - 50° = x° = \underline{40°}$ |

**My Math Examples:**

See students' examples.

# Lesson 8 Triangles

## English Learner Instructional Strategy

### Sensory Support: Mnemonics/Illustrations

Invite volunteers to draw an example of the following angles: *acute angle, right angle, obtuse angle* and label them. Model these mnemonics for *acute* and *obtuse*. Write and say: *The acute angle is a cute little angle.* Discuss how *little/small* things are often described as *cute.* Tell students this can help them remember that acute angles are *smaller* than right angles and *obtuse* angles. Write obtuse and pronounce it, exaggerating the sound of its first syllable by opening your mouth wide and saying, *Ahhhb-tuse.* Then write and say, *Open wide to say obtuse.* This can help them recall that obtuse angles are wider than other angles.

Draw line segments on each angle to make them a closed triangle. Ask students to define the new shapes. **triangles** Continue with the New Vocabulary Activity.

### English Language Development Leveled Activities

| Emerging Level | Expanding Level | Bridging Level |
|---|---|---|
| **Academic Vocabulary** Draw an angle, point to its vertex, and say, *This figure has one **vertex***. Write *vertex*, and have students repeat chorally. Then draw a triangle. Count *1, 2, 3,* as you point to each vertex. Then say, *This figure has three **vertices***. Write *vertices*, and have students repeat chorally. Point to *vertex* and *vertices* on the board, and say them with students again. Then label the triangle's vertices: *J, K, L.* Say, *The vertices are J, K, and L.* Point to each vertex as you name it. Have students use this sentence frame to identify other triangles' vertices: **The vertices are ____.** | **Building Oral Language** Draw a right triangle, an acute triangle, and an obtuse triangle. Point to the right triangle's right angle and say, *One **right** angle; this is a **right** triangle.* Have students repeat chorally. Then point to the acute triangle's 3 acute angles and say, *Three **acute** angles; this is an **acute** triangle.* Have students repeat chorally. Finally, point to the obtuse triangle's obtuse angle and say, *One **obtuse** angle; this is an **obtuse** triangle.* Have students repeat chorally. Then ask them to identify each triangle using this sentence frame: **The triangle is ____ because ____.** | **Shape Riddles** Have students work in groups of three. Direct each group to write a riddle for each type of triangle: acute, right and obtuse. Tell students they must use these terms in their riddles: *angle(s), acute, obtuse, right, perpendicular, line segments,* and *sides.* Have groups solve each other's riddles. |

## Teacher Notes:

NAME _____ DATE _____

# Lesson 8 Vocabulary Definition Map
## *Triangles*

Use the definition map to write a description and list characteristics about the vocabulary word or phrase. Write or draw math examples. Share your examples with a classmate.

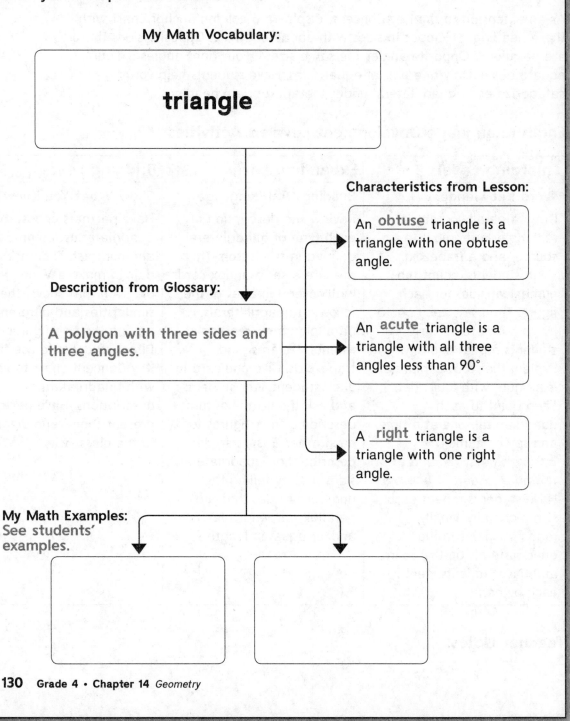

**My Math Vocabulary:**

**triangle**

**Characteristics from Lesson:**

An <u>obtuse</u> triangle is a triangle with one obtuse angle.

**Description from Glossary:**

A polygon with three sides and three angles.

An <u>acute</u> triangle is a triangle with all three angles less than 90°.

A <u>right</u> triangle is a triangle with one right angle.

**My Math Examples:**
See students'
examples.

**130**   Grade 4 · Chapter 14 *Geometry*

# Lesson 9 Quadrilaterals

## English Learner Instructional Strategy

### Vocabulary Support: Anchor Chart

Before the lesson, write the New Vocabulary and Spanish cognates on a cognate chart. Introduce each word, and draw visual math examples. Then discuss how the words in these pairs are related: *parallel/parallelogram, quad/quadrilateral*.

As an extension to the Key Concept, display a 6-column anchor chart with these headings: 1. Opposite sides with equal lengths 2. Opposite sides that are parallel 3. Opposite angles the same size 4. Four right angles 5. Four equal sides 6. Only one pair of equal sides. Have students help you categorize each quadrilateral under the appropriate headings.

### English Language Development Leveled Activities

| Emerging Level | Expanding Level | Bridging Level |
|---|---|---|
| **Word Knowledge** | **Building Oral Language** | **Show What You Know** |
| Draw a parallelogram, a rectangle, a rhombus, a square, and a trapezoid. With students, count the number of sides for each figure. Then say, *Each figure is a quadrilateral.* Have students repeat chorally. Explain that a *quadrilateral* is a figure with four sides. Then point to each quadrilateral, one at a time, and say its name: *parallelogram, rectangle, rhombus, square, trapezoid.* Have students repeat each name chorally. Finally, randomly point to the quadrilaterals on the board, and have students identify each by name. | Review the description of each type of quadrilateral defined in the lesson. Then prepare a set of index cards with several images of the following quadrilaterals: parallelograms, rectangles, rhombi, squares, and trapezoids. Give one card to each student. Pair students and ask them to take turns describing their figures to each other. Ensure students describe their quadrilaterals' sides, i.e., by telling the number of sides that are parallel, perpendicular, and/or equal in length. | Have partners choose two quadrilaterals to compare and contrast. Then ask each pair to make a Venn diagram that shows the similarities and differences among the two figures. Direct students to use the Key Concept chart to assist with quadrilaterals descriptions. Have partners present their Venn diagram to the class. |

### Teacher Notes:

NAME _____ DATE _____

# Lesson 9 Vocabulary Chart
## *Quadrilaterals*

Use the three-column chart to organize the vocabulary in this lesson.
Write the word in Spanish. Then write the correct terms to complete
each definition.

| English | Spanish | Definition |
|---------|---------|------------|
| **parallelogram** | paralelogramo | A quadrilateral in which each pair of opposite sides are __parallel__ and equal in length. |
| **rectangle** | rectángulo | A quadrilateral with _four_ right angles; opposite sides are equal and _parallel_. |
| **rhombus** | rombo | A parallelogram with _four_ congruent sides. |
| **trapezoid** | trapecio | A quadrilateral with exactly _one_ pair of _parallel_ sides. |
| **square** | cuadrado | A rectangle with _four_ congruent sides. |

# Lesson 10 Draw Lines of Symmetry
## English Learner Instructional Strategy

### Sensory Support: Drawings/Models

Draw the following figures on the board:

Point to the line on the first square. Say, *This is a **line of symmetry**.* Model how it divides the square exactly in half by folding a square piece of paper to show 2 equal parts. Then point to the line on the second square. Say, *This is **not** a line of symmetry.* Model how it divides the square into unequal parts by folding a square piece of paper identical to the figure. Distribute 4 square pieces of paper to each student and have them model the remaining 4 squares on the board. Discuss whether each figure is or is not an example of line symmetry.

### English Language Development Leveled Activities

| Emerging Level | Expanding Level | Bridging Level |
|---|---|---|
| **Academic Vocabulary**<br><br>Cut a heart shape from paper by folding the paper in half to ensure that the sides are symmetrical. Display the heart, and point to each of its halves. Say, *The sides are the same. This shape has **line symmetry**.* Have students repeat chorally. Then fold the heart in half and open it again. Run your finger down the crease and say, *This is the line **of** symmetry.* Have students repeat chorally. Repeat the exercise with other symmetrical shapes, such as a butterfly, a square, and a star. | **Listen and Identify**<br><br>Gather and display a mix of five or six symmetrical and asymmetrical shapes. First, point to a symmetrical shape and say, *This shape has **line** symmetry.* Draw a line through the image to divide it in half, and say, *This is the line **of** symmetry.* Next, point to an asymmetrical shape and show how no line can divide it into equal parts. Finally, ask students to identify remaining shapes as you point to them, using this sentence frame: **The shape [does/does not] have line symmetry.** For symmetrical shapes, have them tell where to draw the line of symmetry. | **Building Oral Language**<br><br>Provide a variety of images from magazines or printed from the Internet. Have multilingual groups discuss whether objects and figures shown have line symmetry. Then ask each group to select one image that is symmetrical and one image that is not symmetrical to present to the class. For each image, tell students to either identify the line of symmetry or explain why the item does not have line symmetry. |

### Teacher Notes:

NAME _____ DATE _____

# Lesson 10 Note Taking

## *Draw Lines of Symmetry*

Read the question. Write words you need help with and research each word. Use your lesson to write your Cornell notes. Write or draw math examples to explain your thinking. Share your examples with a classmate.

| | |
|---|---|
| **Building on the Essential Question** <br><br> How do you draw lines of symmetry? | **Notes:** <br><br> A figure has <u>line symmetry</u> if it can be folded so that the two parts of the figure match, or are congruent. <br><br> A <u>line</u> of <u>symmetry</u> is a line on which a figure can be folded so that its two halves match exactly. <br><br> 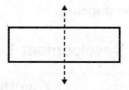 <br><br> Figures **can** have more than one <u>line</u> of <u>symmetry</u>. <br><br> 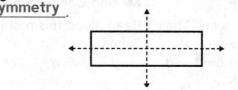 <br><br> Figures can have <u>zero</u> lines of symmetry. <br><br> 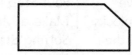 |
| **Words I need help with:** <br> See students' words. | |
| **My Math Examples:** <br><br> See students' examples. | |

# Lesson 11 Problem-Solving Investigation
## Strategy: Make a Model
### English Learner Instructional Strategy

**Collaborative Support: Think-Pair-Share**

Before the lesson, pair emerging students with expanding or bridging students. As you work through the lesson and seek student responses, direct your questions or prompts to student pairs instead of individual students. Give pairs time to think about and discuss their response. Allow the more proficient English speaker to answer. Record his or her answer on the board. Model saying it again and have the class chorally repeat. Be sure to prompt a response from each pair at least once during the lesson.

Then define these words from this lesson's word problems: *chore(s), allowance, washes dishes, walks dog, folds laundry, cycle, photographs, photographed, park, owners, floor, kitchen, tiles.* Use photos, realia, and demonstrations to support students' understanding.

### English Language Development Leveled Activities

| Emerging Level | Expanding Level | Bridging Level |
|---|---|---|
| **Word Knowledge** | **Making Connections** | **Deductive Reasoning** |
| Write 23 on the board. Then show students how to represent the number using base-ten blocks. Say, *My model shows the number 23.* Next, distribute base-ten blocks to students. Have each student use the blocks to model a number of their choice. Then ask each student to present his or her model using this sentence frame: **My model shows the number ____.** Pair emerging students with bridging students to practice the make a model strategy with the riddle activity. | Read aloud this problem: *A triangle has sides of equal length and angles of equal measure. How many lines of symmetry does it have?* Guide students in using the Four-Step Plan and the Make a Model Problem-Solving Strategy to find that the triangle has 3 lines of symmetry. | Have students write a riddle about a shape that does not say what the shape is. As an example, read aloud the following riddle: *I have three sides. Two of my sides are the same length, and they meet at a right angle. What am I?* **a right triangle** After students have written their riddles, ask them to exchange papers with a partner and draw a model to solve each other's riddle. |

**Teacher Notes:**

NAME _____ DATE _____

# Lesson 11 Problem-Solving Investigation
## STRATEGY: Make a Model

**Make a model to solve each problem.**

1. **Mary Anne** is making a pattern with quadrilaterals.
   **She** put **squares** in the <u>first</u> row, **parallelograms** in the <u>second</u> row, and **trapezoids** in the <u>third</u> row.
   She **repeats** this **pattern four** times.
   Which **quadrilateral** does she use in the <u>tenth</u> row?

| Understand | Solve |
|---|---|
| | |
| **Plan**<br><br>I will draw the patterns to make a <u>  model  </u>. | **Check** |

2. Draw <u>two</u> **lines** on the square so that <u>three</u> **right triangles** are formed.

| Understand | Solve |
|---|---|
| I know:<br><br>A right triangle has <u> one </u> right angle.<br><br>A square has <u> four </u> right angles. | |
| **Plan** | **Check** |

**Grade 4 • Chapter 14** *Geometry* **133**

# Dinah Zike Explaining
# Visual Kinesthetic Vocabulary®, or VKVs®

## What are VKVs and who needs them?

❝ VKVs are flashcards that animate words by kinesthetically focusing on their structure, use, and meaning. VKVs are beneficial not only to students learning the specialized vocabulary of a content area, but also to students learning the vocabulary of a second language. ❞

**Dinah Zike | Educational Consultant**

Dinah-Might Activities, Inc. – San Antonio, Texas

## Why did you invent VKVs?

❝ Twenty years ago, I began designing flashcards that would accomplish the same thing with academic vocabulary and cognates that Foldables® do with general information, concepts, and ideas—make them a visual, kinesthetic, and memorable experience. ❞

Dinah Zike's Visual Kinesthetic Vocabulary®

**I had three goals in mind:**

- **Making two-dimensional flashcards three-dimensional**

- **Designing flashcards that allow one or more parts of a word or phrase to be manipulated and changed to form numerous terms based upon a commonality**

- **Using one sheet or strip of paper to make purposefully shaped flashcards that were neither glued nor stapled, but could be folded to the same height, making them easy to stack and store**

## Why are VKVs important in today's classroom?

❝ At the beginning of this century, research and reports indicated the importance of vocabulary to overall academic achievement. This research resulted in a more comprehensive teaching of academic vocabulary and a focus on the use of cognates to help students learn a second language. Teachers know the importance of using a variety of strategies to teach vocabulary to a diverse population of students. VKVs function as one of those strategies. ❞

# Dinah Zike Explaining
# Visual Kinesthetic Vocabulary®, or VKVs®

## How are VKVs used to teach content vocabulary to EL students?

❝ VKVs can be used to show the similarities between cognates in Spanish and English. For example, by folding and unfolding specially designed VKVs, students can experience English terms in one color and Spanish in a second color on the same flashcard while noting the similarities in their roots. ❞

## What organization and usage hints would you give teachers using VKVs?

❝ Cut off the flap of a 6" x 9" envelope and slightly widen the envelope's opening by cutting away a shallow V or half circle on one side only. Glue the non-cut side of the envelope into the front or back of student workbooks or journals. VKVs can be stored in the pocket.

Encourage students to individualize their flashcards by writing notes, sketching diagrams, recording examples, forming plurals (radius: radii or radiuses), and noting when the math terms presented are homophones (sine/sign) or contain root words or combining forms (kilo-, milli-, tri-).

As students make and use the flashcards included in this text, they will learn how to design their own VKVs. Provide time for students to design, create, and share their flashcards with classmates. ❞

Dinah Zike's book Foldables, Notebook Foldables, & VKVs for Spelling and Vocabulary 4th-12th won a Teachers' Choice Award in 2011 for "instructional value, ease of use, quality, and innovation"; it has become a popular methods resource for teaching and learning vocabulary.

Dinah Zike's
**VKV**
Visual
Kinesthetic
Vocabulary ®

Chapter 1

✁ cut on all dashed lines          fold on all solid lines

A digit is (Un dígito es) _____

What is the value of 7 in the number below? (¿Cuál es el valor de 7 en el número de abajo?)

3,758,241

Use commas to separate the periods in the number below. (Usa comas para separar los períodos en el número de abajo.)

3 2 5 4 8 2 2

**digit**

**place value**

**period**

A digit in each place represents _____ times what it would represent in the place to its right. (Un dígito en cada posición representa _____ veces lo que representaría en la posición de su derecha.)

Dinah Zike's
Visual
Kinesthetic
Vocabulary®

Chapter 1

✄ cut on all dashed lines

⬜ fold on all solid lines

posicional

valor

Write the number below in standard form. Then circle the millions period and underline the thousands period. (Escribe el número de abajo en forma estándar. Luego, encierra en un círculo el período de los millones y subraya el período de los millares.)

*Twenty-two million, five hundred sixty-three thousand*

The 2 is in the _____ place. (El 2 está en la posición de _____.)

The 4 has a value of 4 × (El 4 tiene un valor de 4 ×)

| Thousands Period | | | Ones Period | | |
|---|---|---|---|---|---|
| hundreds | tens | ones | hundreds | tens | ones |
| 4 | 6 | 2 | 1 | 5 | 3 |

Write the highlighted digit's place and value. (Escribe el valor y la posición del dígito destacado.)

10,312 _____       154,065 _____

892 _____       9,473 _____

Dinah Zike's
Visual
Kinesthetic
Vocabulary ®

Which shows standard form? (¿Cuál muestra la forma estándar?)

1. 5,000 + 200 + 10 + 6

2. *five thousand, two hundred sixteen*

3. 5,216

On a number line, numbers to the right are _____ than numbers to the left. (En una recta numérica, los números a la derecha son _____ que los números a la izquierda.)

Describe how to round a number. (Describe cómo redondear un número.)

## standard form

## number line

## round

Place a comma correctly into the number below. (Pon una coma correctamente en el número de abajo.)

2 5 4 6 5 8

On a number line, numbers to the left are _____ than numbers to the right. (En una recta numérica, los números a la izquierda son _____ que los números a la derecha.)

Rounding is one way to _____ an answer. (Redondear es una manera de _____ una respuesta.)

Dinah Zike's
Visual
Kinesthetic
Vocabulary®

Chapter 1

✂ cut on all dashed lines

fold on all solid lines

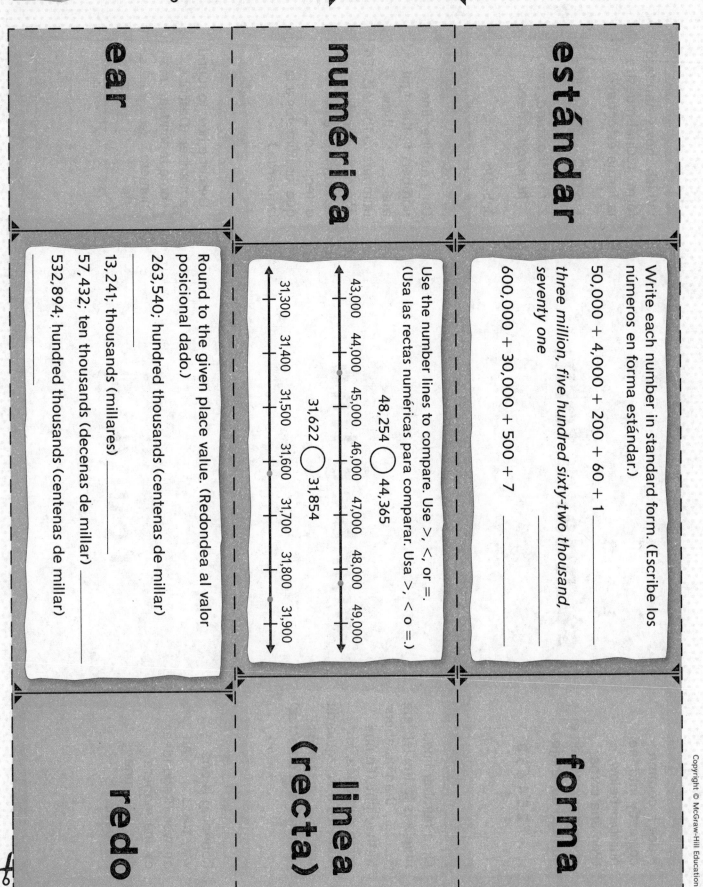

**ear**

**numérica**

**estándar**

Write each number in standard form. (Escribe los números en forma estándar.)

50,000 + 4,000 + 200 + 60 + 1

three million, five hundred sixty-two thousand, seventy one

_____

600,000 + 30,000 + 500 + 7

_____

Use the number lines to compare. Use >, <, or =. (Usa las rectas numéricas para comparar. Usa >, < o =.)

48,254 ◯ 44,365

43,000  44,000  45,000  46,000  47,000  48,000  49,000

31,622 ◯ 31,854

31,300  31,400  31,500  31,600  31,700  31,800  31,900

Round to the given place value. (Redondea al valor posicional dado.)

263,540; hundred thousands (centenas de millar) _____

13,241; thousands (millares) _____

57,432; ten thousands (decenas de millar) _____

532,894; hundred thousands (centenas de millar) _____

**redo**

**línea
(recta)**

**forma**

**Dinah Zike's**
**Visual**
**Kinesthetic**
**Vocabulary** ®

Associative Property of Addition

conmutativa de la suma

Which example shows the Associative Property of Addition? Circle the answer. (¿Qué ejemplo muestra la propiedad asociativa de la suma? Encierra en un círculo la respuesta.)

**1.** $3 + (7 + 5) = (3 + 7) + 5$

**2.** $6 + 4 = 10$ and (y) $4 + 6 = 10$

**3.** $5 + 0 = 5$

Dinah Zike's
**V K V** Visual
Kinesthetic
Vocabulary®

Chapter 2

✂ cut on all dashed lines    ◻ fold on all solid lines

propiedad asociativa de la suma

Commutative

Which example shows the Commutative Property of Addition? Circle the answer. (¿Qué ejemplo muestra la propiedad conmutativa de la suma? Encierra en un círculo la respuesta.)

1. $6 + (3 + 2) = (6 + 3) + 2$

2. $3 + 9 = 12$ and (y) $9 + 3 = 12$

3. $7 + 0 = 7$

Dinah Zike's
**VKV**
Visual
Kinesthetic
Vocabulary ®

Chapter 2

✂ cut on all dashed lines          ⬜ fold on all solid lines

Regroup 12 hundreds. (Reagrupa 12 centenas.)

——— thousand +
——— hundreds

The subtrahend is (El sustraendo es)

———————
———————

Is 255 + 415 an equation? Why or why not? (¿Es 255 + 415 una ecuación? ¿Por qué?)

———————
———————

## regroup

## subtrahend

## equation

Regroup 14 tens. (Reagrupa 14 decenas.)

——— hundred +
——— tens

Circle the subtrahend in the equation below. (Encierra en un círculo el sustraendo en la ecuación de abajo.)

$$
\begin{array}{r}
5,326 \\
-2,114 \\
\hline
3,212
\end{array}
$$

An equation is a type of (Una ecuación es un tipo de) ———————
———————

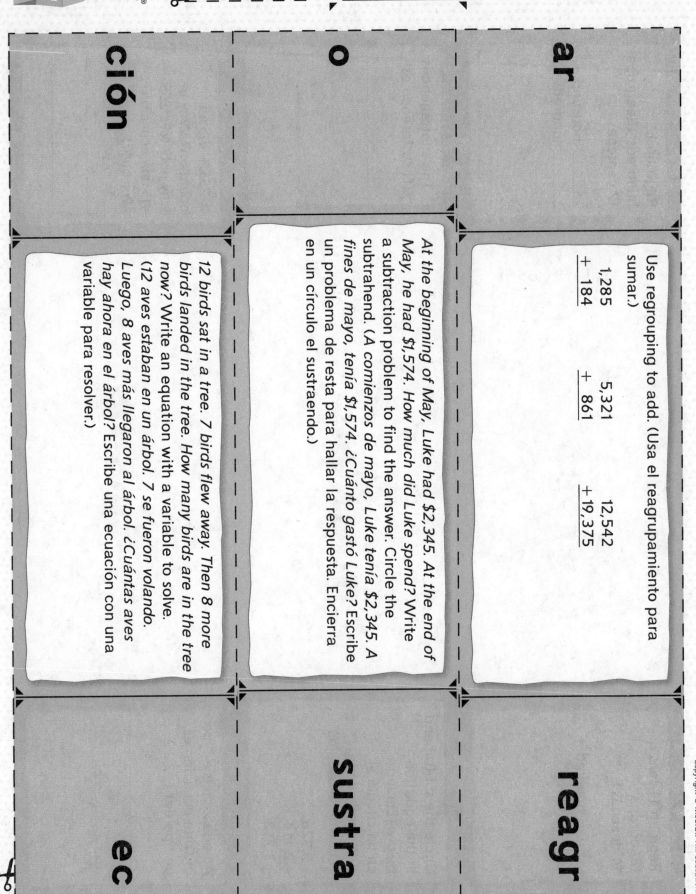

**ción**

**o**

**ar**

Use regrouping to add. (Usa el reagrupamiento para sumar.)

| | | |
|---|---|---|
| 1,285 | 5,321 | 12,542 |
| + 184 | + 861 | + 19,375 |

At the beginning of May, Luke had $2,345. At the end of May, he had $1,574. How much did Luke spend? Write a subtraction problem to find the answer. Circle the subtrahend. (A comienzos de mayo, Luke tenía $1,574. ¿Cuánto gastó Luke? A fines de mayo, tenía $1,574. ¿Cuánto gastó Luke? Escribe un problema de resta para hallar la respuesta. Encierra en un círculo el sustraendo.)

12 birds sat in a tree. 7 birds flew away. Then 8 more birds landed in the tree. How many birds are in the tree now? Write an equation with a variable to solve. (12 aves estaban en un árbol. 7 se fueron volando. Luego, 8 aves más llegaron al árbol. ¿Cuántas aves hay ahora en el árbol? Escribe una ecuación con una variable para resolver.)

**ec**

**sustra**

**reagr**

Dinah Zike's
**VKV** Visual
Kinesthetic
Vocabulary ®

Chapter 3

✂ cut on all dashed lines

▭ fold on all solid lines

Write a fact family for 10, 12, 120. Circle the quotients. (Escribe una familia de operaciones para 10, 12 y 120. Encierra en un círculo los cocientes.)

Circle the factors in each number sentence. (Encierra en un círculo los factores en cada oración numérica.)

$9 \times 5 = 45$

$12 \times 12 = 144$

$6 \times 3 = 18$

Label each part of the multiplication sentence. (Rotula cada parte de la multiplicación.)

$6 \times 9 = 54$

# quotient

# factors

# multiplication

Write a fact family for 7, 8, 56. Circle the quotients. (Escribe una familia de operaciones para 7, 8 y 56. Encierra en un círculo los cocientes.)

Dinah Zike's
Visual
Kinesthetic
Vocabulary®

Chapter 3

✂ cut on all dashed lines

fold on all solid lines

ción

es

e

When you write a multiplication sentence related to a division sentence, will the quotient become a factor or product? (Cuando escribes una multiplicación relacionada a una división, ¿se convierte el cociente en factor o en producto?)

Find each unknown to complete the fact family. Circle the factors. (Halla las incógnitas para completar la familia de operaciones. Encierra en un círculo los factores.)

12 × _____ = 72        _____ × 4 = 28

6 × 12 = _____        _____ × 7 = 28

_____ ÷ 6 = 12        _____ ÷ 7 = 4

72 ÷ _____ = 6        28 ÷ 4 = _____

Hector dealt 7 playing cards to 4 friends. How many cards did he deal altogether? Write a multiplication sentence to solve. (Héctor repartió 7 tarjetas de un juego a 4 amigos. ¿Cuántas tarjetas repartió en total? Escribe una multiplicación para resolver.)

coc

Dinah Zike's
**VKV** Visual
Kinesthetic
Vocabulary ®

✂ cut on all dashed lines       ☐ fold on all solid lines

propiedad conmutativa

Associative Property of Multiplication

Use the Associative Property of Multiplication to show another way to solve. (Usa la propiedad asociativa de la multiplicación para mostrar otra forma de resolver.)

$2 \times 5 \times 7 = 2 \times (5 \times 7)$     $2 \times 5 \times 7 =$ _____

$= 2 \times 35$                              $=$ _____

$= 70$                                       $=$ _____

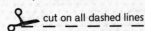

✂ cut on all dashed lines

📄 fold on all solid lines

Which example shows the Commutative Property of Multiplication? Circle the answer. (¿Qué ejemplos muestran la propiedad conmutativa de la multiplicación? Encierra en un círculo la respuesta.)

1. $12 \times 5 = 5 \times 12$

2. $18 \times 1 = 18$

3. $9 \times 0 = 0$

propiedad asociativa de la multiplicación

Commutative Property

Chapter 6

cut on all dashed lines

fold on all solid lines

Dinah Zike's
Visual
Kinesthetic
Vocabulary®

compatible numbers

partial quotients

List 3 partial quotients to help solve $848 \div 4$. (Menciona 3 cocientes parciales que te ayuden a resolver $848 \div 4$.)

Compatible numbers are (Los números compatibles son)

What is the nonmath meaning of *partial*? (¿Cuál es el significado no matemático de *parcial*?)

Dinah Zike's
Visual
Kinesthetic
Vocabulary®

Chapter 6

✂ cut on all dashed lines

fold on all solid lines

**parciales**

**s**

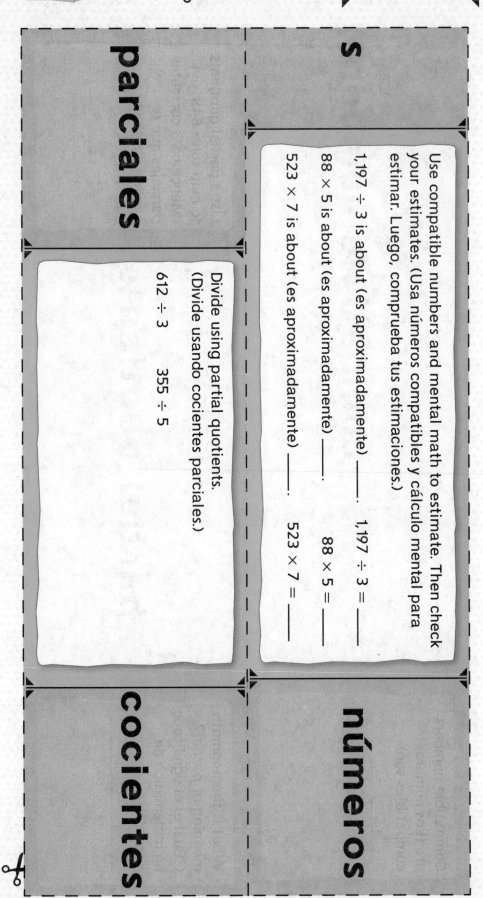

Use compatible numbers and mental math to estimate. Then check your estimates. (Usa números compatibles y cálculo mental para estimar. Luego, comprueba tus estimaciones.)

$1,197 \div 3$ is about (es aproximadamente) _____ . $1,197 \div 3 =$ _____

$88 \times 5$ is about (es aproximadamente) _____ . $88 \times 5 =$ _____

$523 \times 7$ is about (es aproximadamente) _____ . $523 \times 7 =$ _____

Divide using partial quotients. (Divide usando cocientes parciales.)

$612 \div 3$     $355 \div 5$

**cocientes**

**números**

Dinah Zike's
**VKV**
Visual Kinesthetic Vocabulary®

Chapter 7

✂ cut on all dashed lines

fold on all solid lines

## numeric pattern

## order of operations

What rule does the pattern follow? (¿Qué regla sigue el patrón?)

12, 17, 22, 27, 32 _____

Find the value of each expression. (Halla el valor de cada expresión.)

(3 × 5) − (6 ÷ 2) = _____

10 + (25 ÷ 5) = _____

Dinah Zike's
Visual
Kinesthetic
Vocabulary®

Chapter 7

✂ cut on all dashed lines

fold on all solid lines

**ciones**

Write 1-3 on the lines to show the correct order of operations. (Escribe 1, 2 y 3 en la líneas para mostrar el orden de las operaciones correcto.)

_____ Add and subtract in order from left to right. (Suma y resta en orden de izquierda a derecha.)

_____ Perform operations in parentheses. (Haz las operaciones que están entre paréntesis.)

_____ Multiply and divide in order from left to right. (Multiplica y divide en orden de izquierda a derecha.)

**o**

Find the unknown in each numeric pattern. (Halla la incógnita en cada patrón numérico.)

8, 12, 13, 17, 18, 22, 23, _____

55, 75, 65, _____, 75, 95, 85, 105

8, _____, 32, 64, 128

**orden de las**

**patrón**

Dinah Zike's
**VKV**
Visual
Kinesthetic
Vocabulary ®

Chapter 8

✂ cut on all dashed lines

▢ fold on all solid lines

List two factor pairs for 12. (Menciona dos pares de factores para 12.)

_____ and (y) _____

_____ and (y) _____

Circle each prime number. (Encierra en un círculo cada número primo.)

3   21   2   35

17   77   14   39

Write a fraction with a denominator of 9 and a numerator of 5. (Escribe una fracción cuyo denominador sea 9 y el numerador sea 5.)

# factor pairs

# prime number

# fraction

Name two nonmath examples of pairs. (Nombra dos ejemplos no matemáticos de pares.)

_____

_____

Define *prime number*. (Define *número primo*.)

_____

_____

_____

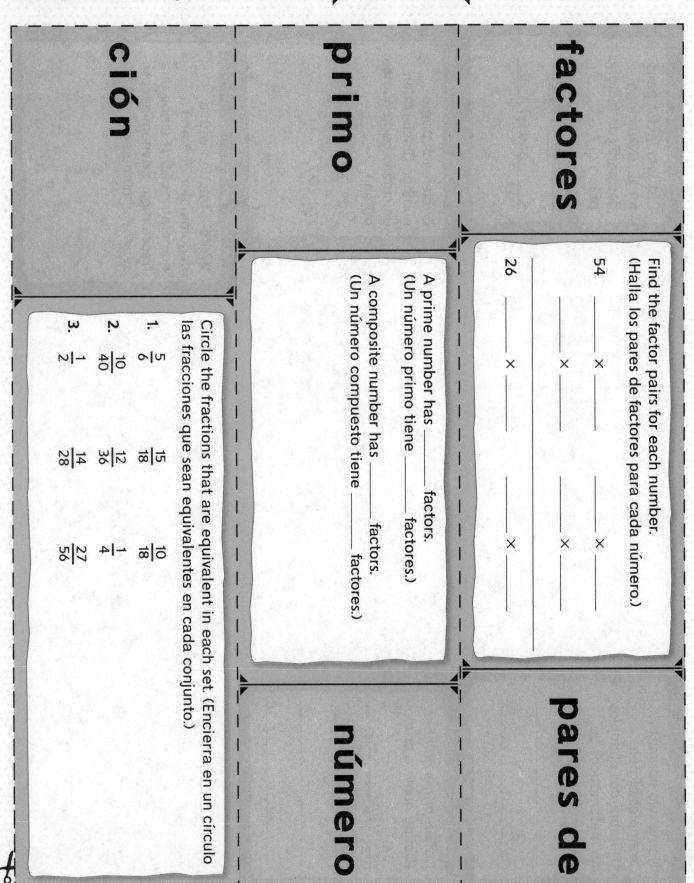

## ción

## primo

## factores

## número

## pares de

**Find the factor pairs for each number.**
(Halla los pares de factores para cada número.)

54 ___ × ___
   ___ × ___
   ___ × ___
   ___ × ___

26 ___ × ___
   ___ × ___

A prime number has _____ factors.
(Un número primo tiene _____ factores.)

A composite number has _____ factors.
(Un número compuesto tiene _____ factores.)

**Circle the fractions that are equivalent in each set.** (Encierra en un círculo las fracciones que sean equivalentes en cada conjunto.)

1. $\frac{5}{6}$    $\frac{15}{18}$    $\frac{10}{18}$

2. $\frac{10}{40}$    $\frac{12}{36}$    $\frac{1}{4}$

3. $\frac{1}{2}$    $\frac{14}{28}$    $\frac{27}{56}$

# least common multiple

# máximo común divisor

Write the greatest common factor (GCF) of the numerator and denominator. Then write the fraction in simplest form. (Escribe el máximo común divisor (M.C.D.) del numerador y el denominador. Luego, escribe la fracción en su mínima expresión.)

$\frac{15}{18}$

$\frac{21}{28}$

GCF: _____

simplest form: _____

GCF: _____

simplest form: _____

Describe why you might need to find the least common multiple of two numbers. (Describe por qué podrías necesitar hallar el mínimo común múltiplo de dos números.)

_____

_____

_____

 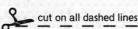
mínimo común múltiplo

List the factors for each number. Circle the greatest common factor. (Menciona los factores de cada número. Encierra en un círculo el máximo común divisor.)

12: ___, ___, ___, ___

18: ___, ___, ___, ___

greatest common factor

Write the least common multiple for each pair of numbers. (Escribe el mínimo común múltiplo de cada par de números.)

5 and (y) 7    2 and (y) 8 _____

3 and (y) 8 _____    2 and (y) 5 _____

Circle the greatest fraction. (Encierra en un círculo la fracción más grande.)

$\frac{3}{5}$    $\frac{12}{25}$    $\frac{1}{2}$

✂ cut on all dashed lines    ✂ fold on all solid lines

Add the unit fractions. (Suma las fracciones unitarias.)

$\frac{1}{5} + \frac{1}{5} + \frac{1}{5} =$ ___

When you simplify, you write a fraction in its ___ form.
(Al simplificar, una fracción se escribe en su ___ expresión.)

Write $\frac{5}{2}$ as a mixed number. (Escribe $\frac{5}{2}$ como número mixto.)

**unit fraction**

**simplify**

**improper fraction**

A unit fraction has a numerator of (Una fracción unitaria tiene un numerador de) ___

Write $6\frac{1}{3}$ as an improper fraction. (Escribe $6\frac{1}{3}$ como fracción impropia.)

Dinah Zike's
Visual
Kinesthetic
Vocabulary®

Chapter 9

✂ cut on all dashed lines    📄 fold on all solid lines

impropia

icar

unitaria

Use the model to find $\frac{1}{8} + \frac{3}{8}$. (Usa el modelo para hallar $\frac{1}{8} + \frac{3}{8}$.)

How many unit fractions did you shade? (Spanish translation) ____

Simplify each fraction. (Simplifica las fracciones.)

$\frac{12}{24} =$ ____    $\frac{4}{16} =$ ____

$\frac{9}{81} =$ ____    $\frac{10}{60} =$ ____

$\frac{24}{36} =$ ____    $\frac{12}{21} =$ ____

Describe how to write a mixed number as an equivalent improper fraction. (Describe cómo escribir un número mixto como una fracción impropia equivalente.)

fracción

fracción

Dinah Zike's
**VKV** Visual
Kinesthetic
Vocabulary®

Chapter 11

✂ cut on all dashed lines          ☐ fold on all solid lines

Use *convert* in a nonmath sentence. (Usa *convertir* en una oración no matemática.)

List two things you would measure in miles. (Menciona dos cosas que medirías en millas.)

Capacity is (La capacidad es)

convert

mile

capacity

Dinah Zike's
Visual
Kinesthetic
Vocabulary ®

Chapter 11

✂ cut on all dashed lines

📄 fold on all solid lines

**dad**

**la**

**ir**

How would you convert 12 yards into feet? (¿Cómo convertirías 12 yardas a pies?)

How would you convert 24 inches into feet? (¿Cómo convertirías 24 pulgadas a pies?)

Mr. Diaz rides his bike 2 miles to work each morning. He rides home again in the afternoon. How far in feet does he ride each day? (El Sr. Díaz anda en bicicleta 2 millas hasta el trabajo todas las mañanas. Vuelve a su casa en bicicleta por la tarde. ¿Cuánto anda en bicicleta cada día?)

1 mile (mi) = 5,280 feet (ft)

Write the units of capacity in order from least to greatest. (Escribe las unidades de capacidad en orden de menor a mayor.)

| quart | cup | ounce | gallon | pint |
| cuarto | taza | onza | galón | pinta |

Dinah Zike's
**VKV**
Visual
Kinesthetic
Vocabulary ®

Chapter 11

✂ cut on all dashed lines

▱ fold on all solid lines

There are ____ quarts in 1 gallon.

(Hay ____ cuartos en 1 galón.)

There are ____ pounds in 2 tons.

(Hay ____ libras en 2 toneladas.)

Write an expression that shows the number of seconds in a day. (Escribe una expresión que muestre el número de segundos que hay en un día.)

____

**quart**

**ton**

**seconds**

There are ____ pints in 1 quart.

(Hay ____ pintas en 1 cuarto.)

Dinah Zike's
**VKV** Visual
Kinesthetic
Vocabulary ®

Chapter 11

✂ cut on all dashed lines

▱ fold on all solid lines

**gundos**

**elada**

**o**

Circle the most reasonable estimate for capacity. (Encierra en un círculo la estimación más razonable de la capacidad.)

| 1 quart | 8 quarts | 25 quarts | 50 quarts |
| (1 cuarto) | 8 cuartos | 25 cuartos | 50 cuartos |

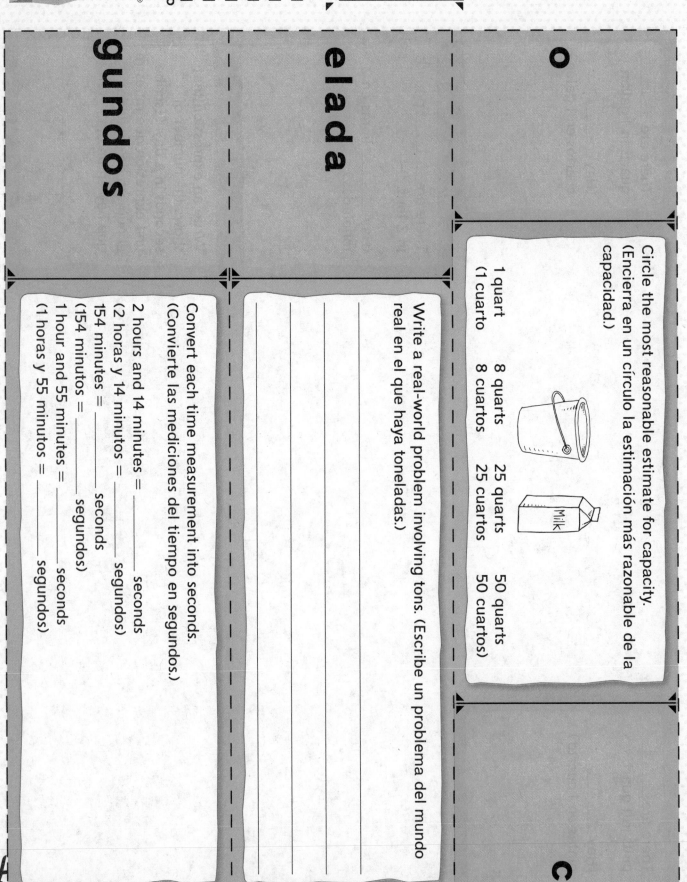

Write a real-world problem involving tons. (Escribe un problema del mundo real en el que haya toneladas.)

Convert each time measurement into seconds. (Convierte las mediciones del tiempo en segundos.)

2 hours and 14 minutes = _____ seconds
(2 horas y 14 minutos = _____ segundos)
154 minutes = _____ seconds
(154 minutos = _____ segundos)
1 hour and 55 minutes = _____ seconds
(1 horas y 55 minutos = _____ segundos)

**c**

_____ mm = 1cm

1m = _____ cm

_____ mL = 1L

How many grams are there in 1 kilogram? (¿Cuántos gramos hay en 1 kilogramo?) _____

**centimeter**

**milliliter**

**kilogram**

Circle the word part in *centimeter* that means "hundred." (Encierra en un círculo la parte de la palabra *centímetro* que significa "cien".)

A milliliter is a _____ unit of _____ capacity.

(Un mililitro es una unidad _____ de capacidad.)

Circle the word part in *kilogram* that means "thousand." (Encierra en un círculo la parte de la palabra *kilogramo* que significa "mil".)

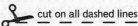

o

ro

ro

Circle the most reasonable estimate for length. (Encierra en un círculo la estimación más razonable de la longitud.)

| | | | |
|---|---|---|---|
| paperclip (clip para papel) | 3 cm | 30 cm | 300 cm |
| book (libro) | 20 cm | 200 cm | 2,000 cm |
| tree (árbol) | 10 cm | 100 cm | 1,000 cm |

Circle the better unit to measure each capacity. (Encierra en un círculo la mejor unidad para medir cada capacidad.)

| | | |
|---|---|---|
| juice box (caja de jugo) | milliliter (mililitro) | liter (litro) |
| fish tank (pecera) | milliliter (mililitro) | liter (litro) |
| ink bottle (tintero) | milliliter (mililitro) | liter (litro) |
| bucket (cubeta) | milliliter (mililitro) | liter (litro) |

List 4 objects that weigh more than 1 kilogram. (Menciona 4 objetos que pesen más de 1 kilogramo.)

_____

_____

kiló

mi

centí

Dinah Zike's
**VKV**
Visual
Kinesthetic
Vocabulary®

Chapter 14

✂ cut on all dashed lines          ▱ fold on all solid lines

Parallel lines are always
(Las líneas paralelas
siempre son)

An obtuse angle
measures (Un
ángulo obtuso mide)
_____ than 180°.
(de 180°.)

What symbol
indicates a right
angle? (¿Qué símbolo
indica un ángulo
recto?)

**parallel**

**obtuse angle**

**right angle**

An obtuse angle
measures (Un
ángulo obtuso mide)
_____ than 90°.
(de 90°.)

A right angle
measures (Un ángulo
recto mide) _____
degrees. (grados.)

Dinah Zike's
Visual
Kinesthetic
Vocabulary®

✂ cut on all dashed lines ⬜ fold on all solid lines

recto

obtuso

ela

Draw a four-sided shape with four right angles.
(Dibuja una figura de cuatro lados con cuatro ángulos rectos.)

Draw two different examples of obtuse angles.
(Dibuja dos ejemplos diferentes de ángulos obtusos.)

Draw an example of the figure.
(Dibuja un ejemplo de la figura.)

$\overline{AB} \parallel \overline{DE}$

ángulo

ángulo

Dinah Zike's
Visual
Kinesthetic
Vocabulary ®

✂ cut on all dashed lines        ⬜ fold on all solid lines

acute triangle

rectángulo

An acute triangle has ——— acute angle(s). Draw
an example of an acute triangle.

(Un triángulo acutángulo tiene ——— ángulo(s)
agudo(s). Dibuja un ejemplo de un triángulo
acutángulo.)

Dinah Zike's
**VKV**
Visual
Kinesthetic
Vocabulary®

Chapter 14

✂ cut on all dashed lines

▭ fold on all solid lines

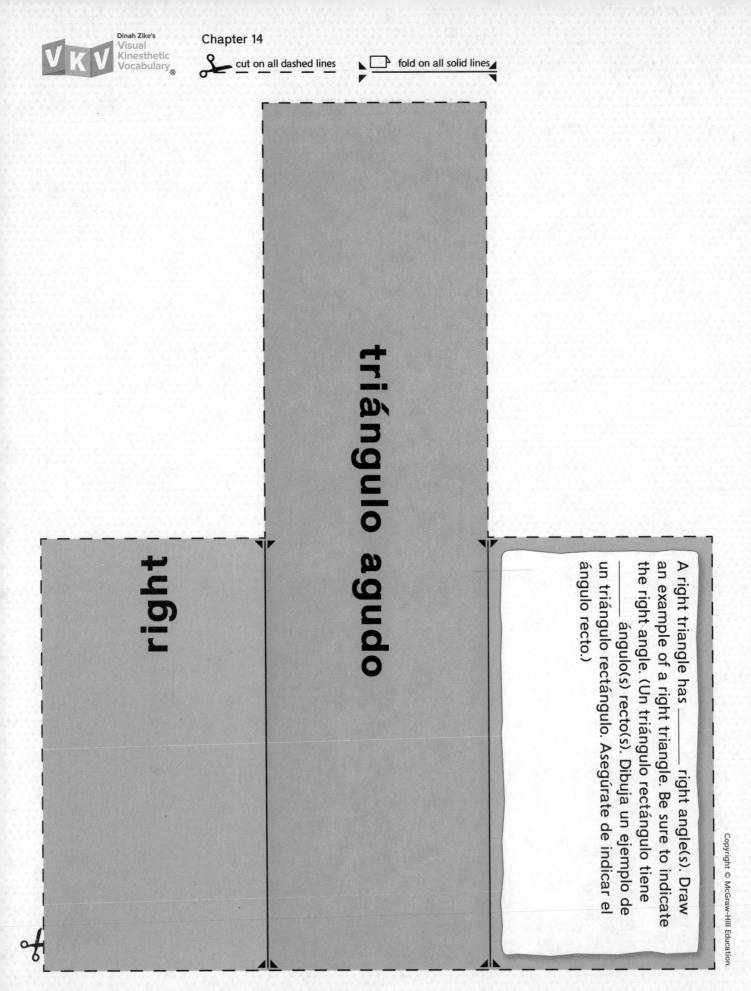

triángulo agudo

right

A right triangle has ———— right angle(s). Draw an example of a right triangle. Be sure to indicate the right angle. (Un triángulo rectángulo tiene ———— ángulo(s) recto(s). Dibuja un ejemplo de un triángulo rectángulo. Asegúrate de indicar el ángulo recto.)

# VKV Answer Appendix

## Chapter 1
### VKV3
*digit:* any symbol used to write whole numbers.
*place value:* 10; seven-hundred thousand
*period:* 3,254,822;

### VKV4
*digito:* hundreds, three hundred; tens, ninety; ten thousands, fifty thousand; thousands, nine thousand
*valor posicional:* thousands; 100,000; 5
*periodo:* 22,563,000

### VKV5
*standard form:* 254,658; 3. 5,216
*number line:* less; greater
*round:* estimate; Sample answer: Look at the number one place to the right of the place you are rounding to; if it is 4 or less, do not change the digit; if it is 5 or greater, increase the digit by one.

### VKV6
*forma estándar:* 54,261; 3,562,071; 630,507
*línea numérica:* >; <
*redondear:* 300,000; 13,000; 60,000; 500,000

## Chapter 2
### VKV7
*Associative Property of Addition:* 1. 3 + (7 + 5) = (3 + 7) + 5

### VKV8
*propiedad asociativa de la suma:* 2. 3 + 9 = 12 and 9 + 3 = 12

### VKV9
*regroup:* 1 hundred + 4 tens; 1 thousand + 2 hundreds
*subtrahend:* 2,114; the number being subtracted
*equation:* number sentence; No, because it does not have an equals sign.

### VKV10
*reagrupar:* See students' work. 1,469; 6,182; 31,917
*sustraendo:* See students' work, $771. $1,574 should be circled.
*ecuación:* Sample answer: $12 - 7 + 8 = x$; $x = 13$

## Chapter 3
### VKV11
*quotient:* $7 \times 8 = 56$, $8 \times 7 = 56$, $56 \div 8 = 7$, $56 \div 7 = 8$; $10 \times 12 = 120$, $12 \times 10 = 120$, $120 \div 10 = 12$, $120 \div 12 = 10$. See students' work for circled quotients.
*factors:* 9 and 5; 12 and 12; 6 and 3
*multiplication:* factor, factor, product

### VKV12
*cociente:* factor
*factores:* 6, 72, 72, 12; 7, 4, 28, 7. See students' work for circled factors.
*multiplicación:* $7 \times 4 = x$; $x = 28$

### VKV13
*Associative Property of Multiplication:* $(2 \times 5) \times 7 = 10 \times 7 = 70$

### VKV14
*propiedad asociativa de la multiplicacion:* 1. $12 \times 5 = 5 \times 12$

# Chapter 6
## VKV15
*compatible numbers:* numbers that are easy to compute mentally.
*partial quotients:* Sample answer: some or part of something; 200, 10, 2

## VKV16
*números compatibles:* 400 and 399; 450 and 440; 3,500 and 3,661
*cocientes parciales:* See students' work.

# Chapter 7
## VKV17
*numeric pattern:* add 5
*order of operations:* 3, 1, 2

## VKV18
*patrón numerico:* 27; 85; 16
*orden de las operaciones:* $15 - 3 = 12$; $10 + 5 = 15$

# Chapter 8
## VKV19
*factor pairs:* Sample answers: glasses and mittens; 2 and 6, 3 and 4
*prime number:* A whole number with exactly two distinct factors, 1 and the number itself; 3, 2, 17
*fraction:* $\frac{5}{9}$

## VKV20
*pares de factores:* $54 \times 1$, $27 \times 2$, $16 \times 3$, $9 \times 6$; $26 \times 1$, $13 \times 2$
*número primo:* 2; 3 or more
*fracción:* $\frac{5}{6}$ and $\frac{15}{18}$; $\frac{10}{40}$ and $\frac{1}{4}$; $\frac{1}{2}$ and $\frac{14}{28}$

## VKV21
*least common multiple:* to create equivalent fractions
*greatest common factor:* 1, 2, 3, 4, 6, 12 and 1, 2, 3, 6, 9, 18; Students should circle both 6s.

## VKV22
*minimo común multiplo:* 35, 24, 8, 10; $\frac{3}{5}$
*máximo común divisor:* 3 and $\frac{5}{6}$; 7 and $\frac{3}{4}$

# Chapter 9
## VKV23
*unit fraction:* 1; $\frac{3}{5}$
*simplify:* simplest
*improper fraction:* $\frac{19}{3}$, $2\frac{1}{2}$

## VKV24
*fracción unitaria:* See students' work; 4
*simplificar:* $\frac{1}{2}$, $\frac{1}{9}$, $\frac{2}{3}$, $\frac{1}{4}$, $\frac{1}{6}$, $\frac{4}{7}$
*fracción impropia:* Sample answer: Multiply the whole number by the denominator of the fraction, then add the result to the numerator.

# Chapter 11
## VKV25
*convert:* Sample answer: I want to convert the shed into a clubhouse.
*mile:* Sample answers: distance between towns; height of a mountain
*capacity:* the amount of liquid a container can hold.

## VKV26
*convertir:* Multiply 12 by 3; divide 24 by 12.
*milla:* $(2 + 2) \times 5,280 = 4 \times 5,280 = 21,120$ ft
*capacidad:* ounce, cup, pint, quart, gallon

## VKV27
*quart:* 2; 4
*ton:* 4,000
*seconds:* $60 \times 60 \times 24$

## VKV28
*cuarto:* 8 quarts; 1 quart
*tonelada:* See students' work.

*segundos:* $(120 \times 60) + (14 \times 60)$
$= 7{,}200 + 840 = 8{,}040;\ 154 \times 60$
$= 9{,}240;\ (60 \times 60) + (55 \times 60) = 3{,}600$
$+ 3{,}300 = 6{,}900$

# Chapter 12
## VKV29
*centimeter:* cent; 10 and 100
*milliliter:* metric; 1,000
*kilogram:* kilo; 1,000

## VKV30
*centímetro:* 3 cm; 20 cm; 1,000 cm
*mililitro:* milliliter; liter; milliliter; liter
*kilógramo: Sample answers:* bowling
ball, car, elephant, bicycle

# Chapter 14
## VKV31
*parallel:* equal distant apart.
*obtuse angle:* more; less
*right angle:* 90; a small square

## VKV32
*paralela:* See students' work.
*ángulo obtuso:* See students' work.
*ángulo recto:* See students' work.

## VKV33
*acute triangle:* three; See students' work.

## VKV34
*triágulo agudo:* one; See students' work.